Studies in Systems, Decision and Control

Volume 239

The series "Studies in Systems, Decision and Control" (SSDC) covers both new developments and advances, as well as the state of the art, in the various areas of broadly perceived systems, decision making and control–quickly, up to date and with a high quality. The intent is to cover the theory, applications, and perspectives on the state of the art and future developments relevant to systems, decision making, control, complex processes and related areas, as embedded in the fields of engineering, computer science, physics, economics, social and life sciences, as well as the paradigms and methodologies behind them. The series contains monographs, textbooks, lecture notes and edited volumes in systems, decision making and control spanning the areas of Cyber-Physical Systems, Autonomous Systems, Sensor Networks, Control Systems, Energy Systems, Automotive Systems, Biological Systems, Vehicular Networking and Connected Vehicles, Aerospace Systems, Automation, Manufacturing, Smart Grids, Nonlinear Systems, Power Systems, Robotics, Social Systems, Economic Systems and other. Of particular value to both the contributors and the readership are the short publication timeframe and the world-wide distribution and exposure which enable both a wide and rapid dissemination of research output.

Indexed by SCOPUS, DBLP, WTI Frankfurt eG, zbMATH, SCImago.

All books published in the series are submitted for consideration in Web of Science.

Libor Pekař

Optimization: An Introduction

Springer

Libor Pekař
Department of Automation and Control
Engineering
Faculty of Applied Informatics
Tomas Bata University in Zlín
Zlín, Czech Republic

Department of Technical Studies
College of Polytechnics Jihlava
Jihlava, Czech Republic

ISSN 2198-4182 ISSN 2198-4190 (electronic)
Studies in Systems, Decision and Control
ISBN 978-3-031-86328-8 ISBN 978-3-031-86326-4 (eBook)
https://doi.org/10.1007/978-3-031-86326-4

This Springer imprint is published by the registered company Springer Nature Switzerland AG
The registered company address is: Gewerbestrasse 11, 6330 Cham, Switzerland

Preface

In various industrial and economic tasks and even in everyday human activity, we face the problem of a solution that depends on a set of unknowns that need to be quantified. Intuitively, the endeavor is to reach a decision that comes the best solution, bringing the highest benefits for us. Such a solution in the form of a set of carefully determined parameters is called optimal.

Parameter optimization deals with finding values of variables that minimize or maximize the so-called objective function. In doing so, there may or may not be ancillary conditions that are called constraints. The application of parameter optimization can be found, for example, in the search for the shortest path, in production planning with an effort to maximize profit and minimize costs, in the planning of business strategies against competitors, in technical practice when minimizing undesirable properties while maintaining the desired properties, in the search for the equilibrium state of a chemical reaction, when solving scientific and technical tasks, etc.

The book intends to bring to the attention essential optimization tools for practitioners and undergraduate students and introduce selected well-established techniques to them when optimizing the parameters of various models. It does not intend to provide the reader with a rigorous mathematic derivation of the presented methods, but it aims to be intelligible for any economically or technically educated person. Each technique is described theoretically and supported by one or more numerical examples that vary from academic ones, through business economics, to funny real-world problems that attract a broad audience. A sketch of Matlab code also follows most of the presented numerical-based techniques.

Readers are mainly provided with a basic introduction to some traditional parameter optimization techniques. The presented problems and their analytic and numerical solution methods represent a core of the parameter optimization reign from the 17th century to the 1970s. Readers are referred to primary literature placed at the end of each chapter to find more detail. Advanced analytic methods based on rigorous mathematics, modern metaheuristic and artificial intelligence methods, the boom we are now experiencing, can be found in other contemporary books and journal articles and are not included in this introductory overview. However, modern techniques often adopt most of the ideas and fundamental principles that readers can find in this book, and they build on them.

I believe that the book finds its place in the libraries of many undergraduate students of various technical study programs and with modern, thoughtful people worldwide, regardless of their expertise.

I am indebted to many people who have helped me immensely in preparing this book. First of all, special thanks go to my wife Eliška and my three beautiful daughters, who provided me with a calm and very pleasant environment for working at home. I am grateful to Prof. Roman Prokop who supervised me on my first teaching and research steps and also showed me the charm of optimization. I am indebted to the Department of Automation and Control Engineering of the Faculty of Applied Informatics, Tomas Bata University in Zlín, and the Department of Technical Studies of the College of Polytechnics Jihlava for providing me with excellent conditions for writing this book. Finally, many inspiring suggestions for improving this material have been given by students in my courses during the last decade.

Zlín/Jihlava, Czech Republic Libor Pekař

Contents

About the Author

Libor Pekař was born in Zlín, Czech Republic, in 1979. He received a B.S. degree in automation and informatics from Tomas Bata University (TBU) in Zlín in 2002, an M.S. degree in automation and control engineering in the consumption industry in 2005, and a Ph.D. degree in technical cybernetics from TBU in Zlín in 2013. He was appointed Associate Professor of machine and process control at TBU in Zlín in 2018.

From 2006 to 2013, he worked as Junior Lecturer at TBU in Zlín. He became Senior Lecturer there in 2013. Since 2020, he has also worked as Senior Lecturer at the College of Polytechnics Jihlava, Czech Republic. In 2024, he was appointed Guest Professor of the Frost Lab at the Beijing Institute of Technology, China.

Assoc. Prof. Pekař is Author of one book and eight book chapters and (Co-) Author of more than 55 journal articles and 80 conference papers. His research interests include analysis, modeling, identification, and control of time-delay systems, algebraic control methods, heat-exchanger processes, autotuning, and optimization techniques. He serves as Associate Editor of *Frontiers in Energy Research* journal, Academic Editor of *Mathematical Problems in Engineering*, Topical Advisory Panel Member and Lead Guest Editor of *Mathematics* journal, Editorial Board Member of *AppliedMath* journal and *Measurement and Control* journal, Advisory Board Member of *Applied Thermal Engineering* journal, and Associate Editor of *International Journal of Robotics and Control Systems* and *Frontiers in Energy Research* journals. He has also served as Reviewer for over 30 scientific journals, including but not limited to *Automatica*, *IEEE Transactions on Automatic Control*, *IEEE Transactions on Industrial Electronics*, *IEEE Control Systems Letters*, *IEEE Access*, and *ISA Transactions*. Assoc. Prof. Pekař received the Laureate of the ASR Seminary Instrumentation and Control in 2007 and 2009, the Rectors' Award for the best Ph.D. thesis at the Faculty of Applied Informatics TBU in Zlín in 2013, Best Reviewer Award from Control Theory and Technology journal in 2019, and the Outstanding Paper Award at the 1st World Conference on Multiphase Transportation, Conversion and Utilization of Energy in 2022.

Chapter 1
Introduction

Optimization means a process of searching for the best solution under given conditions. From the viewpoint of this introductory book, it is considered a mathematical discipline that searches for an **extremum** (i.e., a minimum or a maximum) $\mathbf{x}^*$ of a given **objective function** $f(\mathbf{x})$ defined on a particular set. Hence, the extremum result has usually the form of a point or a vector over real numbers. This kind of optimization is also called parameter optimization, static optimization, or operations research. It is worth noting that this book does not intend to present more advanced problems of the calculus of variations, in which the goal is to find an optimal function $f^*(\mathbf{x})$ that extremizes a certain functional $\Phi(f(\mathbf{x}))$ under specific conditions. The following two simple examples elucidate the difference.

Example 1.1 Elisa's grandmother was asked by the mayor of a small village to knead nut and vanilla doughs for a local folk festival. The mayor promised to refund the grandmother €5 for a kilogram of the nut dough and €4 for a kilogram of the vanilla dough. To prepare a kilogram of the nut dough, one needs 0.4 kg of plain flour. One kilogram of the vanilla dough includes 0.3 kg of flour. However, she has found out that she has only 5 kg of flour at her disposal. The goal is to maximize her profit.

The optimization problem can be formulated as follows:

$$\begin{aligned} \mathbf{x}^* &= \arg\max_{\mathbf{x}} f(\mathbf{x}) = 5x_1 + 4x_2 \\ s.t.: &\quad 0.4x_1 + 0.3x_2 \leq 5 \end{aligned} \tag{1.1}$$

where $\mathbf{x} = (x_1, x_2)$ is a vector of unknown kilograms of nut and vanilla doughs, respectively.

Example 1.2 Consider a famous brachistochrone problem (Fig. 1.1) formulated by Johann Bernoulli (1667–1748). The goal is to find an optimal trajectory $y^*(x)$ of a bead sliding from the initial point A to the final point B, accelerated by the gravitational force F_G without friction, so that the travel time is minimized. By

L. Pekař, *Optimization: An Introduction*, Studies in Systems, Decision and Control 239,
https://doi.org/10.1007/978-3-031-86326-4_1

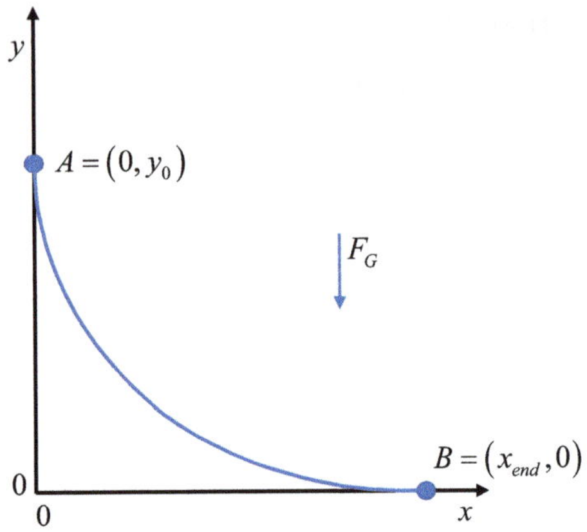

Fig. 1.1 The brachistochrone problem

applying Galileo's law of fall and Pythagoras' theorem, the following optimization problem from the reign of the calculus of variations can be formulated

$$
\begin{aligned}
y^*(x) &= \arg\min_{y(x)} t(y(x)) = \int_0^{x_{end}} \sqrt{\frac{1 + \frac{dy(x)}{dx}}{2gy(x)}} dx \\
s.t.: \quad & y(0) = y_0 \\
& y(x_{end}) = 0
\end{aligned}
\tag{1.2}
$$

where g is the gravitational constant. Readers are referred, e.g., to Boyer et al. [2] for more detail.

Solution methods to the parameter optimization problems presented in this introductory book cover selected analytic and numerical techniques from the nascence of differential calculus to the early works on iterative direct search methods game theory published in decades following World War II. It covers analytic methods to solve one-dimensional and multi-dimensional problems with or without possible constraints.

First, analytic methods to solve one-dimensional and multi-dimensional problems with or without possible constraints are presented. This class of methods is based on differential calculus, investigated mainly by Isaak Newton (1643–1727), Gottfried Wilhelm von Leibniz (1646–1716), and Guillaume de l'Hôpital (1661–1704). For the solution of a constrained problem formulated by a set of equalities, a famous result by Joseph-Louis Lagrange (1736–1813) is used. A more involved task of inequality constraints is solved via the Karush–Kuhn–Tucker conditions, named after William Karush (1917–1997), Harold William Kuhn (1925–2014), and Albert William Tucker (1905–1995).

Second, iterative numerical techniques based on the gradient calculation or its estimation follow. The steepest (gradient) descent, being the leading idea of a broad family of these methods, can be traced back to works by Augustin Louis

Cauchy (1789–1857). For multiple-parameter problems, presented methods vary from a simple strategy by Carl Friedrich Gauss (1777–1855) and Philipp Ludwig von Seidel (1821–1896) to advanced Newton-based (quasi-Newton) computational algorithms developed, e.g., by William Cooper Davidon (1927–2013), Roger Fletcher (1939–2016), and Michael James David Powell (1936–2015).

Readers are also acquainted with gradient-free (or direct-search, comparative) methods. A nice technique based on a famous sequence of numbers introduced to the Western world by Leonardo of Pisa (1170–1240/50), also known as Fibonacci, is presented for constrained one-dimensional problems, whereas an algorithm published by John Nelder (1924–2010) and Roger Mead (1938–2015) crowns a concise overview of comparative methods for unconstrained multi-dimensional parameter optimization tasks.

For constrained problems, general frameworks of penalty and barrier functions are provided to readers. It is worth noting that iterative numerical techniques are usually effortless to be programmed using any procedural programming language. For clarity, solutions of selected tasks are presented in two-dimensional space.

Third, a specific problem of optimizing a linear objective function subject to linear constraints (qualities and/or inequalities) can be solved via linear programming methods. Namely, a solution using the simplex (table) method invented by George Bernard Dantzig (1914–2005) is presented in detail. For an integer programming problem, as a subset of linear programming in which all variables are restricted to be integers, readers are provided with two solution techniques. Namely, the cutting plane method, also known as Gomory's cut, named after Ralph Edward Gomory (1929), and a branch-and-bound algorithm, the framework of which has been proposed by Ailsa Horton Land (1927–2021) and Alison Grant Harcourt (1929). Tasks leading to the family of linear programming can be found mainly in economy, logistics, and production planning.

Fourth, two particular tasks and corresponding solution methods from the reign of dynamic programming are presented. A semi-analytic table-based method optimizing a specifically constraint nonlinear objective function represents one of them. It considers a discrete solution space. Another technique searches for the shortest path between nodes in a weighted connected (possibly oriented) graph. It is also known as Dijkstra's algorithm, named after Dutch scientist Edsger Wybe Dijkstra (1930–2002). Note that dynamic programming is based on general principles formulated by distinguished scholar Richard Ernest Bellman (1920–1984).

Last, this book provides readers with a concise introduction to decision and game theory. The latter is represented by a two-player zero-sum game in the book, expressing an antagonistic conflict. The optimal solution to this kind of problem is based on the Nash equilibrium formulated by John Forbes Nash, Jr. (1928–2015), who was biographed in the book and movie A Beautiful Mind.

The book holding in your hands does not intend in any way to even touch the brilliancy and genius of the men and women mentioned above. The author's intention is to acquaint readers with the rudiments of parameter optimization without overemphasizing math so that the book is readable also to non-experts and a broad audience with general engineering knowledge and to undergraduate students. The aim is not to provide the reader with the most recent findings and methods, the validity and applicability of which are often questionable and unproven. On the contrary, the goal is to present well-established techniques that have been used for decades or even centuries in the reign of parameter optimization. Moreover, practical implementation is emphasized, supported by many examples (approximately half of them are motivated by real-world problems). It, however, does not mean that technical correctness is neglected; the author attempts to avoid factual errors while reducing to necessary math.

With deep respect to books, e.g., by Bertsekas [1], Lange [7], or Rockafellar [8], their rigor, correctness, and profundity make their excellent books understandable only for experts with perfect mathematical comprehension. This book can more likely compete with those intended primarily for students, engineers, practitioners, and researchers who want to use basic ideas of parameter optimization in applied fields like engineering, technology, economics, or others. For instance, kind readers can be referred to titles by Chong et al. [3], Guenin et al. [6], Fischetti [4], French [5], Snyman and Wilke [9], or Winston [10]. This book should serve as an alternate text and a supplement to these excellent introductory and practically-oriented books. It supplements these sources from a different perspective based on ten years of experience from the author's lectures.

References

1. Bertsekas, D.P.: Constrained Optimization and Lagrange Multiplier Methods. Athena Scientific, Nashua (1996)
2. Boyer, C.B., Merzbach, U.C., Asimov, I.: A History of Mathematics, 3rd edn. Wiley, New York (2010)
3. Chong, E.K.P., Lu, W.S., Zak, S.H.: An Introduction to Optimization, 5th edn. Wiley, New York (2023)
4. Fischetti, M.: Introduction to Mathematical Optimization. Independently Published (2019)
5. French, M.: Fundamentals of Optimization: Methods, Minimum Principles, and Applications for Making Things Better. Springer, Cham (2018)
6. Guenin, B., Könemann, J., Tunçel, L.: A Gentle Introduction to Optimization. Cambridge University Press, Cambridge (2014)
7. Lange, K.: Optimization, 2nd edn. Springer, New York (2013)
8. Rockafellar, R.T.: Convex Analysis. Princeton University Press, New Jersey (1997)
9. Snyman, J.A., Wilke, D.N.: Practical Mathematical Optimization: Basic Optimization Theory and Gradient-Based Algorithms 2. Springer, Cham (2019)
10. Winston, W.L.: Operations Research: Applications and Algorithms. Brooks/Cole—Thomson Learning, Belmond (2004)

Chapter 2
Analytic Methods

Let us call methods based on differential calculus the analytic methods. Sometimes, applying these methods is called nonlinear programming, as they can find the minima or maxima of a nonlinear objective function. Solutions to one-dimensional unconstrained problems and multi-dimensional unconstrained and constrained problems are provided to readers in this chapter. Equality and inequality constraints are presented separately since both problems require different techniques and approaches. Readers are referred to [1–8,10–13, 15] for more details.

2.1 One-Dimensional Unconstrained Problem

Before the one-dimensional unconstrained optimization problem is defined and its analytic solution is presented, basic notions and definitions are introduced. It is supposed that readers have elementary knowledge of real-valued functions $f(x) \in \mathbb{R}$ with a real variable $x \in \mathbb{R}$.

2.1.1 Basic Notions

Consider function $f : \mathbb{R} \mapsto \mathbb{R}$ and define the following notions.

Definition 2.1 Function $f(x)$ is **continuous** on interval $I \subseteq \mathbb{R}$ if for every point $x_0 \in I$, it holds that

$$\lim_{x \to x_0} f(x) = f(x_0) \tag{2.1}$$

i.e.,

L. Pekař, *Optimization: An Introduction*, Studies in Systems, Decision and Control 239,
https://doi.org/10.1007/978-3-031-86326-4_2

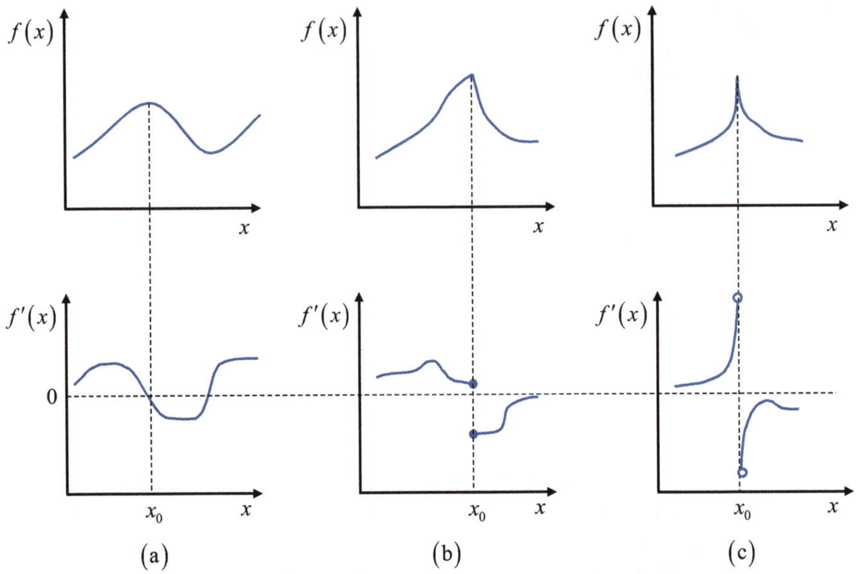

Fig. 2.1 A smooth function (**a**) versus nonsmooth functions (**b**) and (**c**)

$$\forall \varepsilon > 0, \quad \exists \delta > 0, \quad \forall |x - x_0| < \delta, \quad |f(x) - f(x_0)| < \varepsilon \tag{2.2}$$

Definition 2.2 Continuous function $f(x)$ is **smooth** on interval $I \subseteq \mathbb{R}$ if for every point $x_0 \in I$, it holds that $f(x)$ is differentiable (up to the maximum order), i.e., it has all possible derivatives.

Figure 2.1 displays a smooth function (a), versus two types of continuous yet nonsmooth functions with discontinuous first derivatives $f'(x) = df(x)/dx$. Whereas the function in Fig. 2.1b has finite yet nonequal left-hand and right-hand derivatives (i.e., $f'(x)$ has discontinuity of the first kind at x_0), the function in Fig. 2.1c has these derivatives infinite (i.e., $f'(x)$ has discontinuity of the second kind at x_0).

Definition 2.3 Let function $f(x)$ be continuous on interval $I \subseteq \mathbb{R}$. Then, if for every triplet of points $x_1 < x_2 < x_3 \in I$, it holds that:

(a) $f(x_2)$ lies under the line connecting coordinates $[x_1, f(x_1)]$ and $[x_3, f(x_3)]$, the function is **convex** on I. That is

$$f(x_2) < f(x_1) + \frac{f(x_3) - f(x_1)}{x_3 - x_1}(x_2 - x_1) \tag{2.3}$$

(b) $f(x_2)$ lies above the line connecting coordinates $[x_1, f(x_1)]$ and $[x_3, f(x_3)]$, the function is **concave** on I.

$$f(x_2) > f(x_1) + \frac{f(x_3) - f(x_1)}{x_3 - x_1}(x_2 - x_1) \tag{2.4}$$

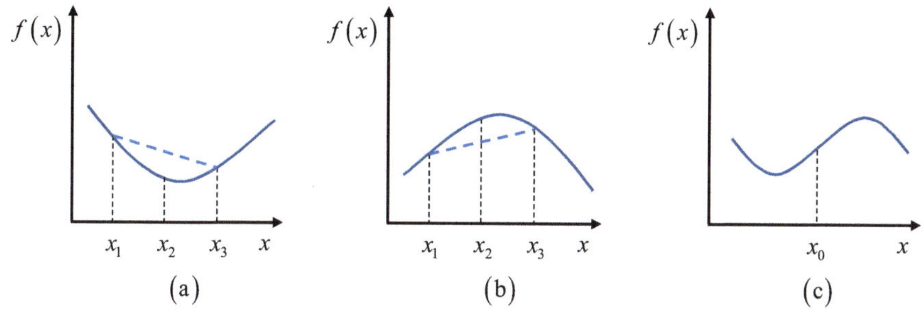

Fig. 2.2 A smooth function (**a**), concave function (**b**), and the point of inflection x_0 (**c**)

Figure 2.2 elucidates these notions. The **point of inflection** is the point at which a smooth plane curve changes from convex to concave (or vice versa).

Definition 2.4 Function $f(x)$ has a **strict local maximum (minimum) point** at x_0 if

$$\exists \delta > 0, \quad \forall x, |x - x_0| < \delta: f(x) < f(x_0) \ (f(x) > f(x_0)) \tag{2.5}$$

If non-strict inequalities ($\leq, \geq$) are used in Definition 2.4, the local extreme point is **non-strict**.

Definition 2.5 Function $f(x)$ has a **global maximum (minimum) point** at x_0 on interval $I \subseteq \mathbb{R}$ if

$$f(x) < f(x_0)(f(x) > f(x_0)) \tag{2.6}$$

holds for all $x \in I$.

Note that the non-strictness can also be considered in the global extremum case. Figure 2.3 displays the meaning of the extrema in Definitions 2.4 and 2.5.

A crucial notion for the extremum computation follows.

Fig. 2.3 On strict and non-strict local and global extrema

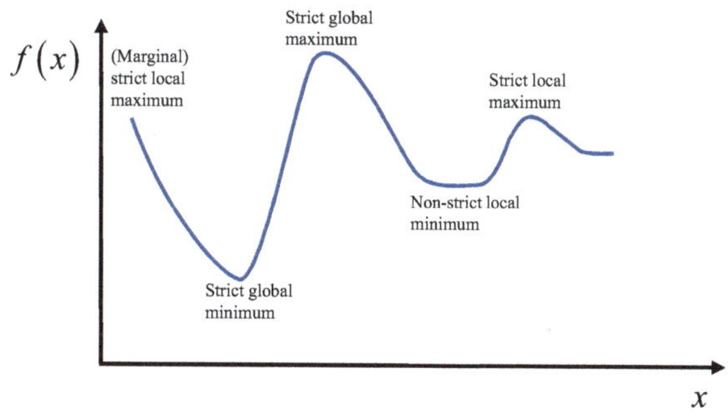

Definition 2.6 Let $f(x)$ be continuous and differentiable (i.e., smooth) on $I \subseteq \mathbb{R}$. If

$$f'(x_0) := \left. \frac{df(x)}{dx} \right|_{x=x_0} = 0 \tag{2.7}$$

then x_0 is called the **stationary point**.

To rephrase Definition 2.6, the stationary point (or points) of $f(x)$ is a value (or, are values) for which the first derivative of the function equals zero.

Definition 2.7 If function $f(x)$ has only one local (or possibly global) extremum on $I \subseteq \mathbb{R}$, it is **unimodal** on I. Otherwise, it is **multimodal**.

2.1.2 *Problem Formulation and Its Solution*

The one-dimensional unconstrained optimization problem means the task of searching a local (or global) maximum and/or minimum of objective function $f(x) \in \mathbb{R}$, i.e.

$$x^* = \arg \operatorname{extr} f(x) \in \mathbb{R}, \; x \in \mathbb{R} \tag{2.8}$$

The following statements show the way to the problem solution. The last one (Theorem 2.1) provides the reader with the eventual solution procedure.

Proposition 2.1 *If $f(x)$ is differentiable at x_0 and has a local maximum or minimum at this point, then $f'(x_0) = 0$ (i.e., x_0 is the stationary point).*

The proposition gives only the necessary (not sufficient) extremum condition. That is, a stationary point might not be an extremum, as sketched in Fig. 2.4. A sufficient condition is provided in Proposition 2.2.

Proposition 2.2 *If x_0 is the stationary point of $f(x)$, and $f'(x_0)$ is increasing (decreasing) at x_0, then $f(x)$ has the strict local minimum (maximum) at x_0.*

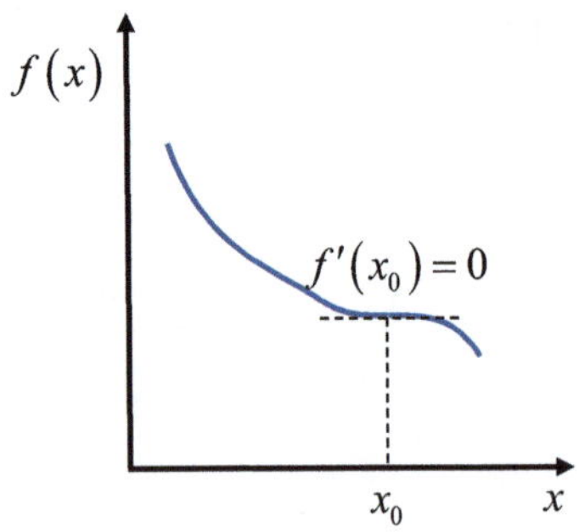

Fig. 2.4 A stationary point not being a local extremum

Hence, the question is how to determine the increase or decrease of the function. In many cases, it is enough to calculate the sign of the second derivative as follows:

- If $f'(x_0) = 0$ and $f''(x_0) > 0$, then $f'(x)$ is increasing at x_0.
- If $f'(x_0) = 0$ and $f''(x_0) < 0$, then $f'(x)$ is decreasing at x_0.
- If $f'(x_0) = 0$ and $f''(x_0) = 0$, then the trend needs to be determined from higher derivatives (if they exist).

Whenever it cannot be decided based on the sign of $f''(x_0)$, the following eventual statement has to be applied that also constitutes the solution guide.

Theorem 2.1 *Let $f(x)$ be differentiable at x_0 up to the order of n and*

$$f(x_0) = f'(x_0) = f''(x_0) = \ldots = f^{(n-1)}(x_0) = 0, f^{(n)}(x_0) \neq 0 \tag{2.9}$$

Then,

(a) *if n is even and $f^{(n)}(x_0) > 0$, then $f(x)$ has a strict local minimum at x_0;*
(b) *if n is even and $f^{(n)}(x_0) < 0$, then $f(x)$ has a strict local maximum at x_0;*
(c) *if n is odd, then $f(x)$ has the point of inflection at x_0.*

To rephrase the theorem, it is necessary to find all stationary points first. Then, do successive derivations until the value of the derivative of $f(x)$ becomes nonzero at each such point. Finally, the decision about the extremum is made based on the derivative order and the sign of the derived function.

Nevertheless, what about points in which the derivative does not exist? For simplicity, consider only continuous functions with a finite number of points in which the function is nonsmooth. The following statement applies to those suspicious points.

Proposition 2.3 *Let $f(x)$ be nonsmooth at x_0. Then*

(a) *if $\lim_{x \to x_0^-} f'(x) > 0$ and $\lim_{x \to x_0^+} f'(x) < 0$, then $f(x)$ has a strict local maximum at x_0;*
(b) *if $\lim_{x \to x_0^-} f'(x) < 0$ and $\lim_{x \to x_0^+} f'(x) > 0$, then $f(x)$ has a strict local minimum at x_0.*

Note that $\lim_{x \to x_0^-} \cdot$ means the left-hand side limit and $\lim_{x \to x_0^+} \cdot$ stands for the right-hand side limit. To rephrase Proposition 2.3, it is necessary to investigate a close neighborhood of each suspicious point. If $f(x)$ decreases in left-hand side neighborhood and increases in the right-hand side neighborhood, there must be a local minimum in the suspicious point. For a local maximum, it holds vice versa.

Note that the same principle can be applied to suspicious points due to discontinuity of $f(x)$.

Several illustrative examples follow.

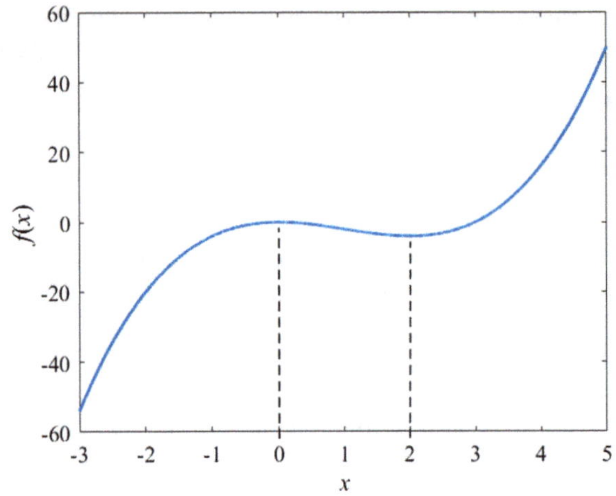

Fig. 2.5 The graph of $f(x) = x^3 - 3x^2$

Example 2.1 *Task*: Find all local extrema of function $f(x) = x^3 - 3x^2$.

Solution: The objective function is differentiable (up to the order of four) on $\mathbb{R}$. Following Theorem 2.1, calculate the first derivative and find stationary points

$$f'(x) = 3x^2 - 6x = 3x(x-2) = 0 \Leftrightarrow x_{0,1} = 0,\ x_{0,2} = 2$$

The second derivative and its value at both stationary points read

$$\begin{aligned} f''(x) &= 6x - 6 = 6(x-1) \\ f''(x_{0,1}) &= f''(0) = -6 \\ f''(x_{0,2}) &= f''(2) = 6 \end{aligned}$$

The second (i.e., even) derivative is nonzero; therefore, there must be extrema in both the points. As $f''(x_{0,1}) < 0$, function $f(x)$ has a strict local maximum at $x = 0$. And $f''(x_{0,2}) > 0$ implies that $f(x)$ has a strict local minimum at $x = 2$, see Fig. 2.5. As the function is smooth and continuous, there are no other local extrema.

Example 2.2 *Task*: Verify that $f(x) = e^x + e^{-x} - x^2$ has a local extremum at $x_0 = 0$.

Solution: Function $f(x)$ is smooth on $\mathbb{R}$. It is necessary to prove that $x_0 = 0$ is the stationary point (according to Proposition 2.1).

$$\begin{aligned} f'(x) &= e^x - e^{-x} - 2x \\ f'(x_0) &= f'(0) = 0 \end{aligned}$$

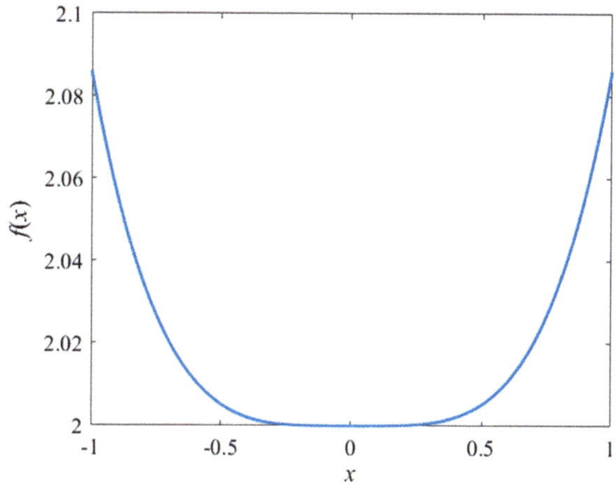

Fig. 2.6 The graph of $f(x) = e^x + e^{-x} - x^2$

Hence, $f(x)$ has the stationary point at $x_0 = 0$. Following Theorem 2.1, calculate higher derivatives

$$
\begin{aligned}
f''(x) &= e^x + e^{-x} - 2 \Rightarrow f''(0) = 0 \\
f'''(x) &= e^x - e^{-x} \Rightarrow f'''(0) = 0 \\
f^{(4)}(x) &= e^x + e^{-x} \Rightarrow f^{(4)}(0) = 2
\end{aligned}
$$

Notice that the existence of a local extremum cannot be verified using Proposition 2.2. As the lowest derivative order giving a nonzero derivative of $f(x)$ equals four (i.e., even) with $f^{(4)}(x_0) > 0$, the function has a strict (yet very flat) local minimum at $x_0 = 0$, see Fig. 2.6.

Example 2.3 *Task*: Find all local extrema of $f(x) = \left|x^2 - x - 2\right|$ on $\mathbb{R}$.

Solution: The objective function is continuous but it is not smooth, as its first derivative does not exist at points $x_{0,1} = -1$ and $x_{0,2} = 2$. Let local extrema be found in $I = \mathbb{R} \backslash \{-1, 2\}$ first.

The objective function can also be represented by

$$
\begin{aligned}
f(x) &= x^2 - x - 2 && \text{for} \quad x \in (-\infty, -1) \cup (2, \infty) \\
f(x) &= -x^2 + x + 2 && \text{for} \quad x \in [-1, 2]
\end{aligned}
$$

The stationary points read

$$
\begin{aligned}
f'(x) &= 2x - 1 \overset{!}{=} 0 \Leftrightarrow x = 0.5 && \text{for} \quad x \in (-\infty, -1) \cup (2, \infty) \\
f'(x) &= -2x + 1 \overset{!}{=} 0 \Leftrightarrow x = 0.5 && \text{for} \quad x \in [-1, 2]
\end{aligned}
$$

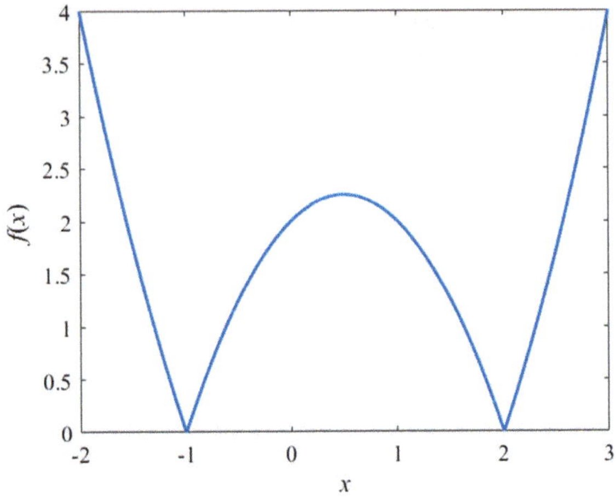

Fig. 2.7 The graph of $f(x) = |x^2 - x - 2|$

where symbol $\stackrel{!}{=}$ means that the left-hand side should be equal to the right-hand side. Hence, $x_0 = 0.5$ is the only stationary point that is, however, valid only for $x \in [-1, 2]$. The decision about a minimum or a maximum can be made based on the second derivative

$$f''(x_0) = f''(0.5) = -2 < 0$$

which implies a strict local maximum at $x_0 = 0.5$ of value $f(0.5) = 2.25$.

The suspicious points $x_{0,1} = -1$ and $x_{0,2} = 2$ can be investigated via Proposition 2.3. Calculate left-hand side and right-hand side limits of first derivatives

$$\lim_{x \to -1^-} f'(x) = -2, \quad \lim_{x \to -1^+} f'(x) = 2$$
$$\lim_{x \to 2^-} f'(x) = -2, \quad \lim_{x \to 2^+} f'(x) = 2$$

It means that $f(x)$ decreases in the left-hand side neighborhood of both suspicious points and increases in the right-hand side neighborhood. Therefore, $x_{0,1} = -1$ and $x_{0,2} = 2$ represent strict local minima of $f(x)$, see Fig. 2.7.

2.2 Multi-dimensional Unconstrained Problem

Most optimization tasks include more than one parameter (variable) to be determined. Hence, the optimal solution is not composed of a single parameter value but a parameter vector. An unconstrained (free) problem considers domain $\mathbb{R}$. If constraints are applied to the domain of the objective function, then a constraint extremum is to be found. Constraints can be expressed by either equalities or inequalities. Whereas the former predominated in history, various problems requiring constrains in the form of inequalities appeared during the twentieth century.

Let basic notions and definitions generalizing the singe-variable case be introduced first. A function $f : \mathbb{R}^n \mapsto \mathbb{R}$ mapping $\mathbf{x} = [x_1, x_2, \ldots, x_n]^T$ to $f(\mathbf{x})$ is considered.

2.2.1 Basic Notions

Definition 2.8 Function $f(\mathbf{x})$ is **convex** if for each pair $\mathbf{x}, \mathbf{y} \in \mathbb{R}^n$, it holds that

$$f(\alpha\mathbf{x} + (1-\alpha)\mathbf{y}) \leq \alpha f(\mathbf{x}) + (1-\alpha)f(\mathbf{y}), \quad 0 \leq \alpha \leq 1 \tag{2.10}$$

The geometrical meaning of convexity is analogous to the one-dimensional case. I.e., the particular function value is lower than the corresponding point lying on the line connecting function values of any two points from the domain. On the contrary, a **concave** function has a function value above the line.

An important feature of convex (concave) functions is that a local minimum (maximum) is also a global one.

Definition 2.9 Let $f(\mathbf{x})$ be continuously differentiable in domain $D \subseteq \mathbb{R}^n$, then the **gradient** of $f(\mathbf{x})$ is the vector of the first partial derivatives

$$\operatorname{grad} f(\mathbf{x}) = \nabla f(\mathbf{x}) := \left[\frac{\partial f(\mathbf{x})}{\partial x_1}, \frac{\partial f(\mathbf{x})}{\partial x_2}, \ldots, \frac{\partial f(\mathbf{x})}{\partial x_n}\right]^T \tag{2.11}$$

In some cases (e.g., if only measured data is available), $f(\mathbf{x})$ is not expressed analytically. Then, $\nabla f(\mathbf{x})$ can be estimated, e.g., using

$$\nabla f(\mathbf{x}_0) \approx [\mathbf{D}(\mathbf{x}_0)]^{-1}\mathbf{d}(f(\mathbf{x}_0)) \tag{2.12}$$

where

$$\mathbf{D}(\mathbf{x}_0) = [\Delta\mathbf{x}_1, \Delta\mathbf{x}_2, \ldots, \Delta\mathbf{x}_n]^T, \ \Delta\mathbf{x}_i = \mathbf{x}_i - \mathbf{x}_0$$
$$\mathbf{d}(f(\mathbf{x}_0)) = \left[\Delta f(\mathbf{x}_1), \Delta f(\mathbf{x}_2), \ldots, \Delta f(\mathbf{x}_n)\right]^T, \ \Delta f(\mathbf{x}_i) = f(\mathbf{x}_i) - f(\mathbf{x}_0)$$

and $\mathbf{x}_i$, $i = 1, 2, \ldots n$ are points from the close neighborhood of $\mathbf{x}_0 \in D$.

The gradient generalizes the first derivative, and indicates the direction of the steepest ascent of $f(\mathbf{x})$ at a particular point, see Fig. 2.8.

Definition 2.10 Let $\nabla f(\mathbf{x})$ be the gradient of $f(\mathbf{x})$ on domain $D \subseteq \mathbb{R}^n$. The **stationary point** $\mathbf{x}_0 \in D$ is a point for which holds that

$$\nabla f(\mathbf{x}_0) := \nabla f(\mathbf{x})|_{\mathbf{x}=\mathbf{x}_0} = \mathbf{0}_{n\times 1} \tag{2.13}$$

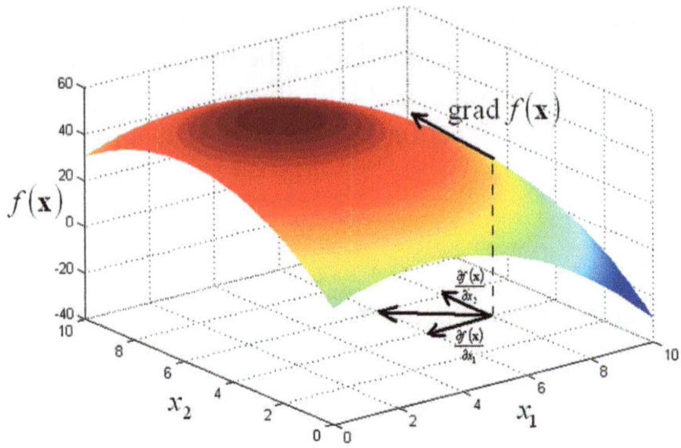

Fig. 2.8 The geometrical meaning of the gradient

where $\mathbf{0}_{n\times 1} = \Bigg[\underbrace{0, 0, \ldots, 0}_{n}\Bigg]^T$. Hence, the stationary point has an analogous meaning to the one-dimensional case.

Definition 2.11 Let $\mathbf{A}$ be a square real-valued matrix of the dimension $n \times n$. The **quadratic form** with matrix $\mathbf{A}$ is function $\varphi(\mathbf{Ax}, \mathbf{x}) = \mathbf{x}^T\mathbf{Ax}$ with $\mathbf{x}_0 \in \mathbb{R}^n$. It is said that the quadratic form is

(a) positive (negative) **semidefinite** if $\varphi(\mathbf{Ax}, \mathbf{x}) \geq 0$ ($\varphi(\mathbf{Ax}, \mathbf{x}) \leq 0$) for every $\mathbf{x} \neq 0 \in \mathbb{R}^n$;
(b) positive (negative) **definite** if $\varphi(\mathbf{Ax}, \mathbf{x}) > 0$ ($\varphi(\mathbf{Ax}, \mathbf{x}) < 0$) for every $\mathbf{x} \neq 0 \in \mathbb{R}^n$;
(c) **indefinite** if its value can be both positive and negative.

It is obvious that the decision about the definiteness cannot be made from Definition 2.11 directly, as it would require testing in finitely many $\mathbf{x}$ vectors. The following proposition provides a useful tool for this decision.

Proposition 2.4 (Sylvester's criterion) *For* $\varphi(\mathbf{Ax}, \mathbf{x})$, *it holds that*

(a) $\varphi(\mathbf{Ax}, \mathbf{x})$ *is positive definite (semidefinite) if and only if all leading principal minors of* $\mathbf{A}$ *are positive (non-negative);*
(b) $\varphi(\mathbf{Ax}, \mathbf{x})$ *is negative definite (semidefinite) if and only if all leading principal minors of* $\mathbf{A}$ *changes their signs starting with negative (nonpositive);*
(c) $\varphi(\mathbf{Ax}, \mathbf{x})$ *is indefinite otherwise.*

The leading principal minors are matrix subdeterminants successively composed by

- the upper left 1×1 corner of $\mathbf{A}$;
- the upper left 2×2 corner of $\mathbf{A}$;
- etc.

Note that the semidefiniteness in items (a) and (b) admits the appearance of the zero(s) in the sequence of the signs.

Definition 2.12 Let $f(\mathbf{x})$ be continuously differentiable up to the order of two in domain $D \subseteq \mathbb{R}^n$. The **Hessian** (or Hesse matrix) $\mathbf{H}(f(x))$ is a symmetric square matrix of the dimension $n \times n$ of the following form

$$\mathbf{H}(f(\mathbf{x})) = \nabla^2 f(\mathbf{x}) := \begin{bmatrix} \frac{\partial^2 f(\mathbf{x})}{\partial x_1^2} & \frac{\partial^2 f(\mathbf{x})}{\partial x_1 \partial x_2} & \cdots & \frac{\partial^2 f(\mathbf{x})}{\partial x_1 \partial x_n} \\ \frac{\partial^2 f(\mathbf{x})}{\partial x_2 \partial x_1} & \frac{\partial^2 f(\mathbf{x})}{\partial x_2^2} & \cdots & \frac{\partial^2 f(\mathbf{x})}{\partial x_2 \partial x_n} \\ \vdots & \vdots & \ddots & \vdots \\ \frac{\partial^2 f(\mathbf{x})}{\partial x_n \partial x_1} & \frac{\partial^2 f(\mathbf{x})}{\partial x_n \partial x_2} & \cdots & \frac{\partial^2 f(\mathbf{x})}{\partial x_n^2} \end{bmatrix} \tag{2.14}$$

The Hessian generalizes the second derivative (i.e., it represents the matrix of all second partial derivatives).

A numerical estimation of $\mathbf{H}(f(x))$ can be computed, e.g., using the following procedure. Let

$$\mathbf{h}_j(f(\mathbf{x}_0)) \approx \frac{\nabla f\left(\mathbf{x}_0 + h\mathbf{e}_j\right) - \nabla f(\mathbf{x}_0)}{h}$$

be the jth column of a Hessian where $\mathbf{e}_j$ is the unit Euclidean vector (with 1 at the jth position), h means a selected discrete (difference) step, and $\mathbf{x}_0$ is the point at which the Hessian is computed. Then, the following formula is used to ensure the Hessian symmetry

$$\tilde{\mathbf{H}}(f(\mathbf{x}_0)) = \frac{\mathbf{H}(f(\mathbf{x}_0)) + \mathbf{H}^T(f(\mathbf{x}_0))}{2}$$

where $\tilde{\mathbf{H}}(f(\mathbf{x}_0))$ expresses the eventual Hessian estimate.

2.2.2 *Problem Formulation and Its Solution*

The multi-dimensional unconstrained optimization problem means the task of searching a local (or global) maximum and/or minimum of objective function $f(\mathbf{x}) \in \mathbb{R}$, $\mathbf{x} \in \mathbb{R}^n$, i.e.

$$\mathbf{x}^* = \arg \operatorname{extr} f(\mathbf{x}) \in \mathbb{R}, \mathbf{x} \in \mathbb{R}^n \tag{2.15}$$

The problem is solved via the following statement providing the *necessary* and *sufficient* local extremum condition.

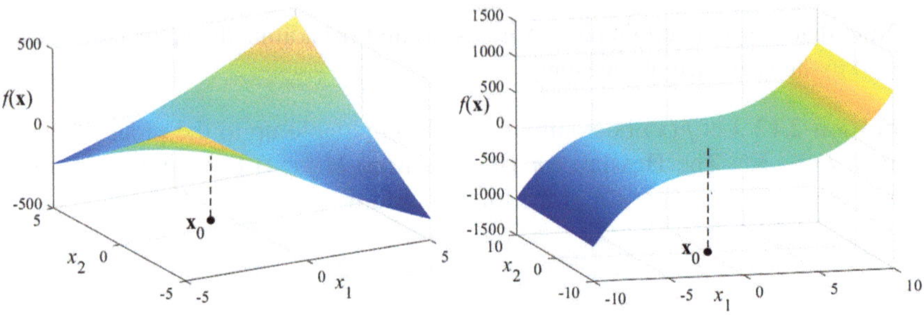

Fig. 2.9 A saddle point (left) and a point of inflection (right) at $\mathbf{x}_0$

Theorem 2.2 *Let $f(\mathbf{x})$ be differentiable up to the order of three at the stationary point $\mathbf{x}_0 \in \mathbb{R}^n$. Then if*

(a) $\varphi(\mathbf{H}(f(\mathbf{x}_0))\mathbf{x}_0, \mathbf{x}_0)$ *is positive definite, then $f(\mathbf{x})$ has a strict local minimum at $\mathbf{x}_0$;*
(b) $\varphi(\mathbf{H}(f(\mathbf{x}_0))\mathbf{x}_0, \mathbf{x}_0)$ *is negative definite, then $f(\mathbf{x})$ has a strict local maximum at $\mathbf{x}_0$;*
(c) $\varphi(\mathbf{H}(f(\mathbf{x}_0))\mathbf{x}_0, \mathbf{x}_0)$ *is indefinite, then $f(\mathbf{x})$ has a either a point of inflection or a saddle point at $\mathbf{x}_0$ (see* Fig. 2.9*);*
(d) $\varphi(\mathbf{H}(f(\mathbf{x}_0))\mathbf{x}_0, \mathbf{x}_0)$ *is semidefinite, then it cannot be decided about the extremum at $\mathbf{x}_0$.*

The saddle point has a local maximum in one direction and a local minimum in another at the same point. Note that the decision about a local extremum as per item (d) of Theorem 2.2 requires further calculations (total differential) beyond this book. Here, the aim is to investigate a local extremum based on items (a) and (b). That is, whenever the quadratic form with the Hessian is definite, the function has a particular local extremum at the stationary point.

In some cases, one can decide about a global extremum, which is formulated in the following proposition.

Proposition 2.5 *If for $\forall \mathbf{x} \in \mathbb{R}^n$, it holds that $\varphi(\mathbf{H}(f(\mathbf{x}))\mathbf{x}, \mathbf{x})$ is positive (negative) semidefinite, the local minimum (maximum) found via* Theorem 2.2 *is also a global minimum (maximum). Moreover, function $f(\mathbf{x})$ is convex (concave).*

Two examples illustrating the use of Theorem 2.2 and Proposition 2.5 follow.

Example 2.4 *Task*: Find all local extrema of $f(\mathbf{x}) = f(x_1, x_2) = -x_1^2 - x_2^2 + x_1x_2 + 2x_1 - x_2 + 100$ on $\mathbb{R}$, and decide whether they are also global extrema.

Solution: The objective function is obviously differentiable on $\mathbb{R}$; hence, calculate all stationary points.

$$\nabla f(\mathbf{x}) = [-2x_1 + x_2 + 2, -2x_2 + x_1 - 1]^T = \mathbf{0}$$

Fig. 2.10 The graph of $f(x_1, x_2) = -x_1^2 - x_2^2 + x_1x_2 + 2x_1 - x_2 + 100$

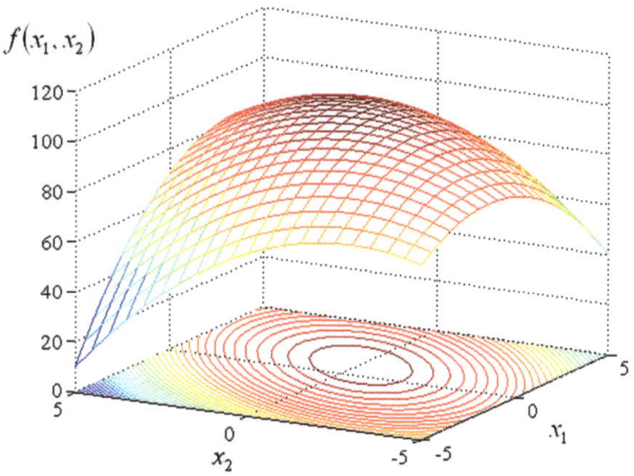

$$\Leftrightarrow x_1 = 1, x_2 = 0$$

Hence, the stationary point is $\mathbf{x}_0 = [1, 0]^T$. The Hessian reads

$$\mathbf{H}(f(\mathbf{x})) = \nabla^2 f(\mathbf{x}) = \begin{bmatrix} -2 & 1 \\ 1 & -2 \end{bmatrix}$$

and its leading principal minors are

$$\left|\nabla^2 f(\mathbf{x})\right|_1 = -2, \; \left|\nabla^2 f(\mathbf{x})\right|_2 = \left|\nabla^2 f(\mathbf{x})\right| = 3$$

According to Proposition 2.4, the quadratic form with the Hessian is negative definite, which implies a strict local maximum at $\mathbf{x}_0$, as per Theorem 2.2 (see Fig. 2.10). Since the Hessian is a constant matrix, its definiteness must hold for all $\mathbf{x} \in \mathbb{R}^2$. Therefore, the local maximum is also global, due to Proposition 2.5.

Example 2.5 *Task*: Find all local extrema of $f(\mathbf{x}) = f(x_1, x_2, x_3) = x_1^3 + x_2^2 + x_3^2 + 12x_1x_2 + 2x_3$ on $\mathbb{R}$, and decide whether they are also global extrema [9].

Solution: The objective function is differentiable on $\mathbb{R}$. The gradient of $f(\mathbf{x})$ reads

$$\begin{aligned} \text{grad} f(\mathbf{x}) = \nabla f(\mathbf{x}) &= \left[\frac{\partial f(\mathbf{x})}{\partial x_1}, \frac{\partial f(\mathbf{x})}{\partial x_2}, \frac{\partial f(\mathbf{x})}{\partial x_3} \right]^T \\ &= \left[3x_1^2 + 12x_2, 12x_1 + 2x_2, 2x_3 + 2\right]^T \end{aligned}$$

The stationary point condition $\nabla f(\mathbf{x}) = \mathbf{0}$ results in the following set of algebraic equations

$$3x_1^2 + 12x_2 = 0$$

$$\begin{aligned} 12x_1 + 2x_2 &= 0 \\ 2x_3 + 2 &= 0 \end{aligned}$$

that has two unambiguous solutions

$$\begin{aligned} \mathbf{x}_{0,1} &= [0, 0, -1]^T \\ \mathbf{x}_{0,2} &= [24, -144, -1]^T \end{aligned}$$

Now, calculate the Hessian and its entries for both stationary points

$$\mathbf{H}(f(\mathbf{x})) = \nabla^2 f(\mathbf{x}) = \begin{bmatrix} \frac{\partial f^2(\mathbf{x})}{\partial x_1^2} & \frac{\partial f^2(\mathbf{x})}{\partial x_1 \partial x_2} & \frac{\partial f^2(\mathbf{x})}{\partial x_1 \partial x_3} \\ \frac{\partial f^2(\mathbf{x})}{\partial x_2 \partial x_1} & \frac{\partial f^2(\mathbf{x})}{\partial x_2^2} & \frac{\partial f^2(\mathbf{x})}{\partial x_2 \partial x_3} \\ \frac{\partial f^2(\mathbf{x})}{\partial x_3 \partial x_1} & \frac{\partial f^2(\mathbf{x})}{\partial x_3 \partial x_2} & \frac{\partial f^2(\mathbf{x})}{\partial x_3^2} \end{bmatrix} = \begin{bmatrix} 6x_1 & 12 & 0 \\ 12 & 2 & 0 \\ 0 & 0 & 2 \end{bmatrix}$$

$$\nabla^2 f\left(\mathbf{x}_{0,1}\right) = \begin{bmatrix} 0 & 12 & 0 \\ 12 & 2 & 0 \\ 0 & 0 & 2 \end{bmatrix}$$

$$\nabla^2 f\left(\mathbf{x}_{0,2}\right) = \begin{bmatrix} 144 & 12 & 0 \\ 12 & 2 & 0 \\ 0 & 0 & 2 \end{bmatrix}$$

Notice that the Hessian symmetry (i.e., $\frac{\partial f^2(\mathbf{x})}{\partial x_i \partial x_j} = \frac{\partial f^2(\mathbf{x})}{\partial x_j \partial x_i}$) may accelerate its calculation.

Following Theorem 2.2, it is necessary to verify the definiteness of quadratic forms with matrices $\nabla^2 f\left(\mathbf{x}_{0,1}\right)$ and $\nabla^2 f\left(\mathbf{x}_{0,2}\right)$. Their leading principal minors (according to the Sylvester's criterion) read

$$\begin{aligned} &\left|\nabla^2 f\left(\mathbf{x}_{0,1}\right)\right|_1 = 0, \ \left|\nabla^2 f\left(\mathbf{x}_{0,1}\right)\right|_2 = \begin{vmatrix} 0 & 12 \\ 12 & 2 \end{vmatrix} = -144, \\ &\left|\nabla^2 f\left(\mathbf{x}_{0,1}\right)\right|_3 = \left|\nabla^2 f\left(\mathbf{x}_{0,1}\right)\right| = -288 \\ &\left|\nabla^2 f\left(\mathbf{x}_{0,2}\right)\right|_1 = 144, \ \left|\nabla^2 f\left(\mathbf{x}_{0,2}\right)\right|_2 = \begin{vmatrix} 144 & 12 \\ 12 & 2 \end{vmatrix} = 144, \\ &\left|\nabla^2 f\left(\mathbf{x}_{0,2}\right)\right|_3 = \left|\nabla^2 f\left(\mathbf{x}_{0,2}\right)\right| = 288 \end{aligned}$$

The quadratic form with $\nabla^2 f\left(\mathbf{x}_{0,1}\right)$ is indefinite; therefore, $f(\mathbf{x})$ has not a local extremum at $\mathbf{x}_{0,1}$. Contrariwise, the quadratic form with $\nabla^2 f\left(\mathbf{x}_{0,2}\right)$ is positive definite, which implies the strict local minimum of $f(\mathbf{x})$ at $\mathbf{x}_{0,2}$.

It remains to be determined whether the local extremum is also global. As the quadratic form with the Hessian is indefinite for $x_1 < 12$, the objective function is not convex (obviously, $\lim_{x_1 \to \infty} f(x_1, \cdot) \to -\infty$), $\mathbf{x}_{0,2}$ is not a global minimum point.

2.3 Multi-dimensional Equality Constrained Problem

Optimization problems formulated by constraints in the form of equalities are considered in this section. A set of these constraints composes the feasible set of candidate solutions, i.e., the admissible domain.

2.3.1 Problem Formulation and Its Solution

The multi-dimensional equality-constrained optimization problem means the task of searching a maximum and/or minimum of objective function $f(\mathbf{x}) \in \mathbb{R}$, $\mathbf{x} \in \mathbb{R}^n$, i.e.

$$\mathbf{x}^* = \arg \operatorname{extr} f(\mathbf{x}) \in \mathbb{R}, \mathbf{x} \in \mathbb{R}^n \tag{2.16}$$

subject to equality constraints

$$g_j(\mathbf{x}) = 0,\ j = 1, \ldots, m \leq n \tag{2.17}$$

Note that whenever the number of independent constraints (m) is higher than the dimension of $\mathbf{x}$ over $\mathbb{R}$ (n), i.e., $m > n$, set (2.17) does not have a feasible solution. If $m = n$, the optimization problem is reduced solely to the solution of (2.17) (if it exists). Then, one can only compare values of $f(\mathbf{x})$ at the solution points; hence, task (2.16) is dull. Therefore, condition $m < n$ is meaningful in practical optimization problems.

Define a crucial notion for constrained problems.

Definition 2.13 The **Lagrangian function** $\Phi(\mathbf{x}, \boldsymbol{\lambda})$ is defined as

$$\Phi(\mathbf{x}, \boldsymbol{\lambda}) = f(\mathbf{x}) + \sum_{j=1}^{m} \lambda_j g_j(\mathbf{x}) \tag{2.18}$$

where $\boldsymbol{\lambda} = [\lambda_1, \lambda_2, \ldots, \lambda_m]^T \in \mathbb{R}^m$ are so-called **Lagrange multipliers.**

The following *necessary* extremum condition provides a core of the problem solution.

Theorem 2.3 (Lagrange multipliers theorem) *Let $f(\mathbf{x})$, $g_1(\mathbf{x})$, $g_2(\mathbf{x})$, ..., $g_m(\mathbf{x})$ be real-valued functions with continuous first partial derivatives. In addition, let gradients $\nabla g_j(\mathbf{x})$, $j = 1, 2, \ldots, m$, be linearly independent functions (i.e., the Jacobian matrix $\mathbf{J}(g_j(\mathbf{x})) = [\nabla g_1(\mathbf{x}), \nabla g_2(\mathbf{x}), \ldots, \nabla g_m(\mathbf{x})]^T \in \mathbb{R}^{m \times n}$ composed by these gradients is of row rank m for all $\mathbf{x} \in \mathbb{R}^n$). Then, if $f(\mathbf{x})$ has a local extremum at $\mathbf{x}_0$ with respect to constraints $g_j(\mathbf{x}) = 0$, $j = 1, 2, \ldots, m$, then there exist $\lambda_1, \lambda_2, \ldots, \lambda_m \in \mathbb{R}$, for which it holds that*

$$\nabla_{\mathbf{x}}\Phi(\mathbf{x}_0, \boldsymbol{\lambda}) = \nabla_{\mathbf{x}} f(\mathbf{x}_0) + \sum_{j=1}^{m} \lambda_j \nabla_{\mathbf{x}} g_j(\mathbf{x}_0) = \mathbf{0}_{n\times 1} \tag{2.19}$$

where $\mathbf{0}_{n\times 1} = \left[\underbrace{0, 0, \ldots, 0}_{n}\right]^T$.

To rephrase Theorem 2.3, the necessary local extremum condition is that all partial derivatives of the Lagrangian function with respect to $\mathbf{x}$ are zero, or equivalently, a local extremum $\mathbf{x}_0$ of $f(\mathbf{x})$ must be the stationary point of $\Phi(\mathbf{x}, \boldsymbol{\lambda})$ (under particular differentiation conditions). Note that constraint conditions (2.17) are also considered when calculating $\mathbf{x}_0$. These conditions coincide with $\nabla_{\boldsymbol{\lambda}}\Phi(\mathbf{x}_0, \boldsymbol{\lambda}) = \mathbf{0}_{m\times 1}$.

Usually, several points satisfying Theorem 2.3 are found. The decision about the sufficient condition(s) is not so straightforward. It holds that the local extremum is a saddle point of the Lagrangian function [14]; see Theorem 2.2. However, it is not easy to distinguish between the saddle point and the point of inflection by calculations. Therefore, other techniques investigating $\mathbf{x}_0$ from (2.19) are used:

- If $f(\mathbf{x})$ is convex (concave) and all constraints in (2.17) are linear, the stationary point $\mathbf{x}_0$ found via Theorem 2.3 is a local and global minimum (maximum).
- If $\mathrm{d}g(\mathbf{x}_0) = \left[\nabla g_1(\mathbf{x}_0), \nabla g_2(\mathbf{x}_0), \ldots, \nabla g_m(\mathbf{x}_0)\right]^T \mathbf{x}_0 = \mathbf{J}\left(g_j(\mathbf{x}_0)\right)\mathbf{x}_0 = \mathbf{0}$ and the quadratic form with the Hessian is positive (negative), then $\mathbf{x}_0$ from Theorem 2.3 is a local minimum (maximum) of $f(\mathbf{x})$ with respect to (2.17).
- Let (2.17) compose a compact (i.e., close and bounded) set. Then, the local maximum *of* $f(\mathbf{x})$ with respect to (2.17) is $\mathbf{x}^*_{\max} = \arg\max f\left(\mathbf{x}_{0,i}\right)$, and the local minimum is $\mathbf{x}^*_{\min} = \arg\min f\left(\mathbf{x}_{0,i}\right)$, where $\mathbf{x}_{0,1}, \mathbf{x}_{0,2}, \ldots$ are solutions of (2.17) and (2.19).

As most equality constraints compose a compact set in practice, the last listed technique is usually used. It means that values of $f(\mathbf{x})$ at all $\mathbf{x}_0$ from (2.17) and (2.19) are simply compared.

In some cases, constraints (2.17) may vanish by substituting variables from them into (2.19), which yields a one-dimensional or multi-dimensional unconstrained problem (see Sects. 2.1 and 2.2).

The meaning and interpretation of the Lagrange multipliers go beyond this introductory book. They express a sensitivity of the extremal $f(\mathbf{x})$ value with respect to an infinitesimal change of constraints (2.17).

Example 2.6 *Task*: Find all extrema of $f(\mathbf{x}) = f(x_1, x_2) = x_1 x_2$ subject to constraint $g(x_1, x_2) = x_1^2 + x_2^2 - 1 = 0$, i.e., a unit circle centered at the origin [9].

Solution: Both $f(\mathbf{x})$ and $g(\mathbf{x})$ have continuous first derivatives. The Lagrangian function reads

$$\Phi(x_1, x_2, \lambda) = f(x_1, x_2) + \lambda g(x_1, x_2) = x_1 x_2 + \lambda\left(x_1^2 + x_2^2 - 1\right)$$

Use Theorem 2.3 to find suspicious extremum points

$$\frac{\partial \Phi(x_1, x_2, \lambda)}{\partial x_1} = x_2 + 2\lambda x_1 \overset{!}{=} 0$$
$$\frac{\partial \Phi(x_1, x_2, \lambda)}{\partial x_2} = x_1 + 2\lambda x_2 \overset{!}{=} 0$$
$$g(x_1, x_2) = x_1^2 + x_2^2 - 1 = 0$$

i.e., these conditions are $\nabla_{\mathbf{x}} \Phi(x_1, x_2, \lambda) = 0$, $g(x_1, x_2) = 0$. The set of three independent algebraic equations has the following four solutions that constitute suspicious points at which the extremum can or cannot be

$$\mathbf{x}_{0,1} = \left[-\frac{\sqrt{2}}{2}, -\frac{\sqrt{2}}{2}\right]^T, \mathbf{x}_{0,2} = \left[-\frac{\sqrt{2}}{2}, \frac{\sqrt{2}}{2}\right]^T,$$
$$\mathbf{x}_{0,3} = \left[\frac{\sqrt{2}}{2}, -\frac{\sqrt{2}}{2}\right]^T, \mathbf{x}_{0,4} = \left[\frac{\sqrt{2}}{2}, \frac{\sqrt{2}}{2}\right]^T$$

Since set $S = \{\mathbf{x} \in \mathbb{R}^2 : g(\mathbf{x}) = 0\}$ is compact, the extremum exists. Now, compare values of $f(\mathbf{x})$ at the suspicious points

$$f(\mathbf{x}_{0,1}) = f(\mathbf{x}_{0,4}) = 0.5, \; f(\mathbf{x}_{0,2}) = f(\mathbf{x}_{0,3}) = -0.5$$

It implies that $f(\mathbf{x})$ has two (strict) local maxima at $\mathbf{x}_{0,1} = \left[-\frac{\sqrt{2}}{2}, -\frac{\sqrt{2}}{2}\right]^T$, $\mathbf{x}_{0,4} = \left[\frac{\sqrt{2}}{2}, \frac{\sqrt{2}}{2}\right]^T$ and two (strict) local minima at $\mathbf{x}_{0,2} = \left[-\frac{\sqrt{2}}{2}, \frac{\sqrt{2}}{2}\right]^T$, $\mathbf{x}_{0,3} = \left[\frac{\sqrt{2}}{2}, -\frac{\sqrt{2}}{2}\right]^T$ with respect to constraint set S.

The geometric interpretation of the given optimization problem is depicted in Fig. 2.11. Hence, the extrema of $f(\mathbf{x})$ are searched on the projection of the unit circle.

2.4 Multi-dimensional Inequality Constrained Problem

Sometimes, optimization problems involve constraints expressed also by inequalities. To illustrate the geometric meaning of inequality constraints by a simple example in $\mathbb{R}^2$, equality $x_1^2 + x_2^2 - 1 = 0$ expresses the unit circle (see Example 2.6), whereas inequality $x_1^2 + x_2^2 - 1 \leq 0$ means a unit disk. These tasks are more complicated than in the equality case and require more advanced solution techniques. Since many tasks require a non-negative solution in practice, the corresponding specified guide for $\mathbf{x} \geq \mathbf{0}_{n\times 1}$ is also given to readers, besides a general solution.

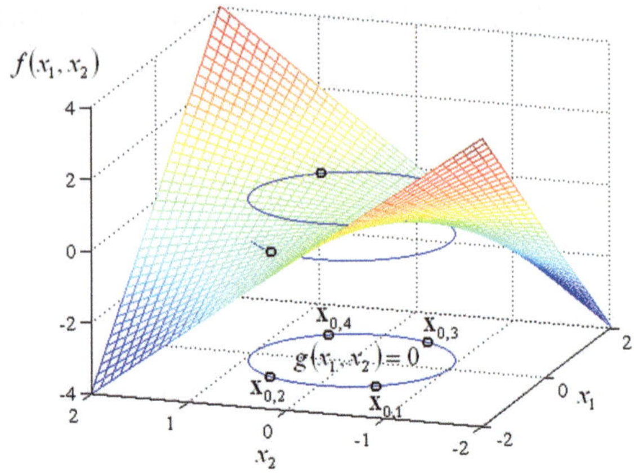

Fig. 2.11 The graph of $f(x_1, x_2) = x_1 x_2$ subject to constraint $g(x_1, x_2) = x_1^2 + x_2^2 - 1 = 0$

2.4.1 General Problem Formulation and Its Solution

Consider the multi-dimensional inequality-constrained optimization problem as the task of searching a minimum of objective function $f(\mathbf{x}) \in \mathbb{R}$, $\mathbf{x} \in \mathbb{R}^n$, i.e.

$$\mathbf{x}^* = \arg\min f(\mathbf{x}) \in \mathbb{R},\ \mathbf{x} \in \mathbb{R}^n \tag{2.20}$$

subject to inequality constraints

$$g_j(\mathbf{x}) \leq 0,\ j = 1, \ldots, m \tag{2.21}$$

Noteworthy, there is no upper limit to m in (2.21), i.e., it is meaningful to consider also $m > n$ if suitable. The problem (2.20)–(2.21) does not require a non-negative solution.

Before the analytic solution is presented, some "tricks" are introduced to cope with tasks that do not satisfy (2.20)–(2.21). Namely, one may require finding a maximum instead of a minimum. Or what should be done if $g_j(\mathbf{x}) \geq 0$ for some j?

- If $\max f(\mathbf{x})$ is required, the problem can be converted to $\min f(\mathbf{x})$ using the following rule

$$\min f(\mathbf{x}) = -\max(-f(\mathbf{x})) \tag{2.22}$$

- If $g_j(\mathbf{x}) \geq 0$ is required, it can simply be multiplied by -1, which yields $-g_j(\mathbf{x}) \leq 0$.

Equation (2.22) means that the original objective function $f(\mathbf{x})$ is multiplied by -1 and the maximum problem becomes the minimum one (say $h(\mathbf{x}) = -f(\mathbf{x})$). Simultaneously, the eventually found optimum point is identical for both the problems, i.e., $\mathbf{x}^* = \arg\max f(\mathbf{x}) = \arg\min h(\mathbf{x})$. However, if one wants to know the original optimum function value $f(\mathbf{x}^*)$, it is necessary to multiply the value by -1 again, i.e., $f(\mathbf{x}^*) = -h(\mathbf{x}^*)$.

To imagine this situation geometrically, consider Fig. 2.10. If the function is multiplied by -1, the maximum peak becomes the minimum one (i.e., the concave function becomes convex); however, the extremum remains at $\mathbf{x}^* = [1, 0]^T$.

The two above-given rules hold generally and hence can be used not only for the analytic solution of the multi-dimensional inequality-constrained optimization problem.

The following analytic Karush–Kuhn–Tucker theorem provides the necessary and sufficient extremum condition.

Theorem 2.4 (Karush–Kuhn–Tucker theorem) *Let* $f(\mathbf{x})$, $g_1(\mathbf{x})$, $g_2(\mathbf{x})$, ... $g_m(\mathbf{x})$ *be real-valued functions and set* $S = \{\mathbf{x} \in \mathbb{R}^n : g_j(\mathbf{x}) = 0, j = 1, 2, \ldots, m\}$ *be convex. Then,* $\mathbf{x}_0 = \mathbf{x}^*$ *is the optimal solution of problem (2.20)–(2.21) if and only if there exists vector* $\boldsymbol{\lambda}^* = [\lambda_1^*, \lambda_2^*, \ldots, \lambda_m^*]^T \geq \mathbf{0}_{m\times 1}$, *so that for every* $\mathbf{x} \geq \mathbf{0}_{n\times 1}$, $\boldsymbol{\lambda} \geq \mathbf{0}_{m\times 1}$, *it holds that*

$$\Phi(\mathbf{x}^*, \boldsymbol{\lambda}) \leq \Phi(\mathbf{x}^*, \boldsymbol{\lambda}^*) \leq \Phi(\mathbf{x}, \boldsymbol{\lambda}^*) \tag{2.23}$$

where $\Phi(\mathbf{x}, \boldsymbol{\lambda})$ *is the Lagrangian function.*

Theorem 2.4 is also called the saddle point theorem as it expresses that the optimal solution of the multi-dimensional inequality constraint problem is represented by a minimum *of* $\Phi(\mathbf{x}, \boldsymbol{\lambda})$ in the $\mathbf{x}$ direction and, simultaneously, a maximum of $\Phi(\mathbf{x}, \boldsymbol{\lambda})$ along $\boldsymbol{\lambda}$. In addition, none of the Lagrange multipliers can be negative.

However, the theorem is hardy usable for an analytic calculation; nevertheless, it can be implemented by numerical (iterative) methods introduced in Chap. 3. In this chapter, the following necessary condition can be used.

Theorem 2.5 (Karush–Kuhn–Tucker local conditions) *Let* $f(\mathbf{x})$, $g_1(\mathbf{x})$, $g_2(\mathbf{x})$, ..., $g_m(\mathbf{x})$ *be real-valued functions with continuous first partial derivatives and* $\mathbf{x}_0 \in \mathbb{R}^n$ *be a local minimum of* $f(\mathbf{x})$ *for problem (2.20)–(2.21). Then, there exists* $\boldsymbol{\lambda}_0 = [\lambda_{0,1}, \lambda_{0,2}, \ldots, \lambda_{0,m}]^T \geq \mathbf{0}_{m\times 1}$ *so that*

$$\begin{aligned}
&\nabla_{\mathbf{x}}\Phi(\mathbf{x}_0, \boldsymbol{\lambda}_0) = \mathbf{0}_{n\times 1} \\
&\nabla_{\boldsymbol{\lambda}}\Phi(\mathbf{x}_0, \boldsymbol{\lambda}_0) \leq \mathbf{0}_{m\times 1} \\
&\lambda_{0,j}\frac{\partial\Phi(\mathbf{x}_0, \boldsymbol{\lambda}_0)}{\partial\lambda_j} = 0, \; j = 1, \ldots, m
\end{aligned} \tag{2.24}$$

Note that the second condition in (2.24) is identical to inequalities (2.21) at $\mathbf{x} = \mathbf{x}_0$, $\boldsymbol{\lambda} = \boldsymbol{\lambda}_0$. Recall that notation $\partial\Phi(\mathbf{x}_0, \boldsymbol{\lambda}_0)/\partial\cdot$ means $[\partial\Phi(\mathbf{x}, \boldsymbol{\lambda})/\partial\cdot]_{\mathbf{x}=\mathbf{x}_0, \boldsymbol{\lambda}=\boldsymbol{\lambda}_0}$.

2.4.2 Non-negative Problem Formulation and Its Solution

Now let the general multi-dimensional inequality constrained optimization problem be supplemented by the non-negativity solution condition $\mathbf{x} \geq \mathbf{0}_{n\times 1}$. As in the general case, this condition may not be restrictive.

- If $x_i < 0$ is required for some $i = 1, 2, \ldots, n$, it can be replaced by introducing two other variables $x_{i1} \geq 0, x_{i2} \geq 0$ and adding the eventual binding condition $x_i = \min(x_{i1}, x_{i2}) - \max(x_{i1}, x_{i2})$.

 Theorem 2.5 then turns into the following statement.

Theorem 2.6 (Karush–Kuhn–Tucker local conditions for a non-negative case) *Let $f(\mathbf{x})$, $g_1(\mathbf{x})$, $g_2(\mathbf{x})$, ..., $g_m(\mathbf{x})$ be real-valued functions with continuous first partial derivatives and $\mathbf{x}_0 = \left[x_{0,1}, x_{0,2}, \ldots, x_{0,n}\right]^T \geq \mathbf{0}_{n\times 1}$ be a local minimum of $f(\mathbf{x})$ for problem (2.20)–(2.21) with $\mathbf{x} \geq \mathbf{0}_{n\times 1}$. Then, there exist $\boldsymbol{\lambda}_0 = \left[\lambda_{0,1}, \lambda_{0,2}, \ldots, \lambda_{0,m}\right]^T \geq \mathbf{0}_{m\times 1}$ so that*

$$\begin{aligned} &\nabla_{\mathbf{x}}\Phi(\mathbf{x}_0, \boldsymbol{\lambda}_0) \geq \mathbf{0}_{n\times 1} \\ &\nabla_{\boldsymbol{\lambda}}\Phi(\mathbf{x}_0, \boldsymbol{\lambda}_0) \leq \mathbf{0}_{m\times 1} \\ &x_{0,i}\frac{\partial\Phi(\mathbf{x}_0, \boldsymbol{\lambda}_0)}{\partial x_i} = 0, \;\; i = 1, \ldots, n \\ &\lambda_{0,j}\frac{\partial\Phi(\mathbf{x}_0, \boldsymbol{\lambda}_0)}{\partial \lambda_j} = 0, \; j = 1, \ldots, m \end{aligned} \tag{2.25}$$

2.4.3 Example

The following simple illustrative example demonstrates the use of both Theorems 2.5 and 2.6.

Example 2.7 *Task*: Find a minimum of $f(\mathbf{x}) = (x_1 - 3)^2 + x_2^2$ subject to constraints

$$\begin{aligned} &2x_1 + x_2 - 2 \leq 0 \\ &x_1, x_2 \geq 0 \end{aligned} \tag{2.26}$$

Solution: Let the given optimization problem be solved in two different ways.

First, let the problem be solved generally via Theorem 2.5. Introduce $g_1(\mathbf{x}) = g_1(x_1, x_2) := 2x_1 + x_2 - 2$, $g_2(\mathbf{x}) = g_2(x_1, x_2) := -x_1$ and $g_3(\mathbf{x}) = g_3(x_1, x_2) := -x_2$, as the non-negativity constraints can also be expressed as $-x_1 \leq 0$ and $-x_2 \leq 0$. Functions $f(\mathbf{x})$ and $g_j(\mathbf{x})$, $j = 1, 2, 3$ have continuous first derivatives. The Lagrangian function reads

$$\Phi(x_1, x_2, \lambda_1, \lambda_2, \lambda_3) = (x_1 - 3)^2 + x_2^2 + \lambda_1(2x_1 + x_2 - 2) + \lambda_2(-x_1) + \lambda_3(-x_2)$$

and the necessary extremum condition (2.24) has the form of a set of the following equalities and inequalities

$$\begin{aligned}
&\nabla_{\mathbf{x}}\Phi(\mathbf{x}, \boldsymbol{\lambda}) = [2(x_1 - 3) + 2\lambda_1 - \lambda_2, 2x_2 + \lambda_1 - \lambda_3]^T = [0, 0]^T \\
&\nabla_{\boldsymbol{\lambda}}\Phi(\mathbf{x}, \boldsymbol{\lambda}) = [2x_1 + x_2 - 2, -x_1, -x_2]^T \leq [0, 0, 0]^T \\
&\lambda_1 \frac{\partial\Phi(\mathbf{x}, \boldsymbol{\lambda})}{\partial\lambda_1} = \lambda_1(2x_1 + x_2 - 2) = 0 \\
&\lambda_2 \frac{\partial\Phi(\mathbf{x}, \boldsymbol{\lambda})}{\partial\lambda_2} = -\lambda_2 x_1 = 0 \\
&\lambda_3 \frac{\partial\Phi(\mathbf{x}, \boldsymbol{\lambda})}{\partial\lambda_3} = -\lambda_3 x_2 = 0 \\
&x_1, x_2, \lambda_1, \lambda_2, \lambda_3 \geq 0
\end{aligned}$$

that can be solved, e.g., starting with the quadruple of equations. These equations have five solutions:

$$\begin{aligned}
\mathbf{x}_{0,1} &= [3, 0]^T, \ \boldsymbol{\lambda}_{0,1} = [0, 0, 0]^T \\
\mathbf{x}_{0,2} &= [0, 0]^T, \ \boldsymbol{\lambda}_{0,2} = [0, -6, 0]^T \\
\mathbf{x}_{0,3} &= [1, 0]^T, \ \boldsymbol{\lambda}_{0,3} = [2, 0, 2]^T \\
\mathbf{x}_{0,4} &= [0, 2]^T, \ \boldsymbol{\lambda}_{0,4} = [-4, -14, 0]^T \\
\mathbf{x}_{0,5} &= [1.4, -0.8]^T, \ \boldsymbol{\lambda}_{0,5} = [1.6, 0, 0]^T
\end{aligned}$$

Apparently, solutions $[\mathbf{x}_{0,2}, \boldsymbol{\lambda}_{0,2}]^T$, $[\mathbf{x}_{0,4}, \boldsymbol{\lambda}_{0,4}]^T$ and $[\mathbf{x}_{0,5}, \boldsymbol{\lambda}_{0,5}]^T$ are not non-negative. Solution $[\mathbf{x}_{0,1}, \boldsymbol{\lambda}_{0,1}]^T$ does not satisfy constraint $2x_1+x_2-2 \leq 0$. Therefore, $\mathbf{x}_{0,3} = [1, 0]^T$, $\boldsymbol{\lambda}_{0,3} = [2, 0, 2]^T$ is the only solution satisfying the set of necessary conditions; hence, $\mathbf{x}_0 = \mathbf{x}_{0,3} = [1, 0]^T$ is a (local) minimum point. This fact is depicted graphically in Fig. 2.12.

Second, let the problem be solved using Theorem 2.6 now. It means that the non-negativity condition is assumed implicitly. Then, only one constraint is introduced, $g(\mathbf{x}) = g(x_1, x_2) := 2x_1 + x_2 - 2$. The corresponding Lagrangian function is

$$\Phi(\mathbf{x}, \lambda) = \Phi(x_1, x_2, \lambda) = (x_1 - 3)^2 + x_2^2 + \lambda(2x_1 + x_2 - 2)$$

The necessary extremum condition set (2.25) reads

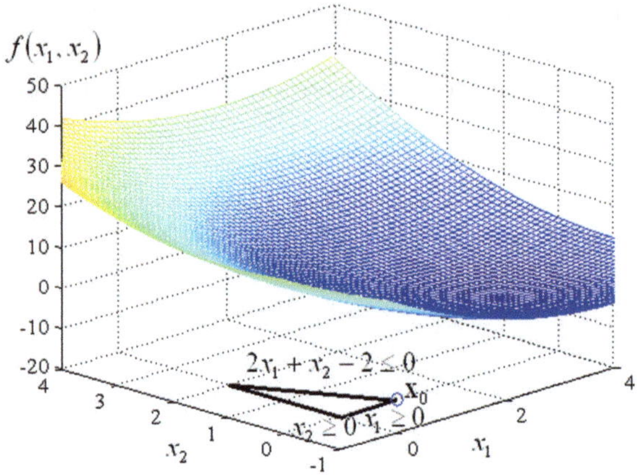

Fig. 2.12 The graph of $f(\mathbf{x}) = (x_1 - 3)^2 + x_2^2$ subject to constraints $2x_1 + x_2 - 2 \le 0, x_1 \ge 0$, and $x_2 \ge 0$

$$\nabla_{\mathbf{x}} \Phi(\mathbf{x}, \lambda) = [2(x_1 - 3) + 2\lambda, 2x_2 + \lambda]^T \ge [0, 0]^T$$
$$\nabla_{\lambda} \Phi(\mathbf{x}, \lambda) = g(x_1, x_2) = 2x_1 + x_2 - 2 \le 0$$
$$x_1 \frac{\partial \Phi(\mathbf{x}, \lambda)}{\partial x_1} = x_1(2(x_1 - 3) + 2\lambda) = 0$$
$$x_2 \frac{\partial \Phi(\mathbf{x}, \lambda)}{\partial x_2} = x_2(2x_2 + \lambda) = 0$$
$$\lambda \frac{\partial \Phi(\mathbf{x}, \lambda)}{\partial \lambda} = \lambda(2x_1 + x_2 - 2) = 0$$
$$x_1, x_2, \lambda \ge 0$$

The triplet of equations has the following five solutions

$$\mathbf{x}_{0,1} = [3, 0]^T, \ \lambda_{0,1} = 0$$
$$\mathbf{x}_{0,2} = [0, 0]^T, \ \lambda_{0,2} = 0$$
$$\mathbf{x}_{0,3} = [1, 0]^T, \ \lambda_{0,3} = 2$$
$$\mathbf{x}_{0,4} = [0, 2]^T, \ \lambda_{0,4} = -4$$
$$\mathbf{x}_{0,5} = [1.4, -0.8]^T, \ \lambda_{0,5} = 1.6$$

Again, solutions $\left[\mathbf{x}_{0,2}, \lambda_{0,2}\right]^T$, $\left[\mathbf{x}_{0,4}, \lambda_{0,4}\right]^T$ and $\left[\mathbf{x}_{0,5}, \lambda_{0,5}\right]^T$ are not non-negative, and $\left[\mathbf{x}_{0,1}, \lambda_{0,1}\right]^T$ does not satisfy $g(x_1, x_2) \le 0$. Therefore, the given objective function subject to constraints (2.26) has a minimum at $\mathbf{x}_0 = \mathbf{x}_{0,3} = [1, 0]^T$ again.

References

1. Bertsekas, D.P.: Constrained Optimization and Lagrange Multiplier Methods. Athena Scientific, Nashua (1996)
2. Chong, E.K.P., Lu, W.S., Zak, S.H.: An Introduction to Optimization, 5th edn. Wiley, New York (2023)
3. Fischetti, M.: Introduction to Mathematical Optimization. Independently Published (2019)
4. Fletcher, R.: Practical Methods of Optimization, 2nd edn. Wiley, New York (1987)
5. French, M.: Fundamentals of Optimization: Methods, Minimum Principles, and Applications for Making Things Better. Springer, Cham (2018)
6. Gill, P.E., Murray, W., Wright, M.H.: Practical Optimization. Academic Press, London (1981)
7. Guenin, B., Könemann, J., Tunçel, L.: A Gentle Introduction to Optimization. Cambridge University Press, Cambridge (2014)
8. Hillier, F.S., Lieberman, G.J.: Introduction to Operations Research, 7th edn. McGraw-Hill, New York (2002)
9. Kopáček, J.: Matematická analýza nejen pro fyziky II (Mathematical Analysis for Physicists II). Matfyzpress Praha, Prague (2015)
10. Lange, K.: Optimization, 2nd edn. Springer, New York (2013)
11. Rardin, R.L.: Optimization in Operations Research, 2nd edn. Pearson, London (2016)
12. Snyman, J.A., Wilke, D.N.: Practical Mathematical Optimization: Basic Optimization Theory and Gradient-Based Algorithms 2. Springer, Cham (2019)
13. Tabak, D., Kuo, B.C.: Optimal Control by Mathematical Programming. Prentice-Hall, New York (1971)
14. Walsh, G.R.: Methods of Optimization. Wiley, New York (1975)
15. Winston, W.L.: Operations Research: Applications and Algorithms. Brooks/Cole—Thomson Learning, Belmond (2004)

Chapter 3
Iterative Methods

Iterative methods are based on a successive improvement of solution estimation. Starting from an initial guess, a sequence of improving approximate solutions is obtained. It means that receiving an unambiguous optimal solution "at once" is impossible, but the sequence of solution guesses can differ for different techniques and their settings. Nevertheless, if the technique suits a particular problem well, the sequence (at least asymptotically) approaches the actual optimum.

Some methods use calculations (or computations) of derivatives, gradients, etc. They are called gradient-based. However, the objective function may not be provided as an analytic expression but only via a sequence of variable values and corresponding function values, or it is not differentiable, or its derivative calculation is demanding. In these cases, the so-called comparative (or direct) methods apply that are based on comparisons of the objective function values without derivative calculations or estimations. They can also be used when a region in which an extremum exists is a priori known.

The general principle of iterative methods for one-dimensional and multi-dimensional problems, respectively, can simply be expressed as

$$
\begin{aligned}
x_{(i+1)} &= x_{(i)} + \Delta_{(i)}, \quad x \in \mathbb{R} \\
\mathbf{x}_{(i+1)} &= \mathbf{x}_{(i)} + \mathbf{\Delta}_{(i)}, \quad \mathbf{x} \in \mathbb{R}^n
\end{aligned}
\tag{3.1}
$$

where subscript (i) means the *iterative step* number, i.e., $\left\{x_{(i)}\right\}_{i=1}^{\infty}$ (or $\left\{\mathbf{x}_{(i)}\right\}_{i=1}^{\infty}$) represents a sequence of optimal solution estimates. An update of the solution guess at each iteration step i, $\Delta_{(i)}$ (or $\mathbf{\Delta}_{(i)}$), depends on a particular solution technique or method. If should hold that

$$
\begin{aligned}
\lim_{i \to \infty} x_{(i)} &= x^* \\
\lim_{i \to \infty} \mathbf{x}_{(i)} &= \mathbf{x}^*
\end{aligned}
\tag{3.2}
$$

L. Pekař, *Optimization: An Introduction*, Studies in Systems, Decision and Control 239,
https://doi.org/10.1007/978-3-031-86326-4_3

where the asterisk denotes the actual optimal solution.

If readers are interested in learning more about iterative optimization methods, the author can refer them to, for example, [1–24].

3.1 Gradient-Free Methods

Three simple methods to one-dimensional problems are introduced first. Then, three multi-dimensional techniques for unconstrained problems follow. All the methods are supported by commented sketches of Matlab codes.

3.1.1 *One-Dimensional Problem*

The unconstrained one-dimensional optimization problem has been defined in (2. 8). However, the presented methods initially consider a closed interval $x \in [a, b]$ representing an infinite set of possible solution candidates. A common principle of all the presented methods lies in a successive reduction of the original interval, until a sufficient accuracy is reached. The reduction is based on a calculation of two additional points inside the current interval followed by a cancelation of the lower or upper bound point.

Recall that the objective function can be represented by a (sufficiently dense) sequence of parameter values $x_1, x_2, \ldots, x_n$ and their corresponding function values $f(x_1), f(x_2), \ldots, f(x_n)$. In this case, the set of possible solutions is finite.

3.1.1.1 Fibonacci Method

Let the goal of to find *a minimum* (without loss of generality) of $f : \mathbb{R} \mapsto \mathbb{R}$ on interval $x \in [a, b]$ of the function domain. The method reduces the current interval based on a ratio of two subsequent values of the so-called Fibonacci sequence.

Definition 3.1 The **Fibonacci sequence** $\{F_k\}_{k=0}^{N}, F_k \in \mathbb{N}$, is generated using the rule

$$\begin{aligned} F_k &= F_{k-1} + F_{k-2} \\ F_0 &= F_1 = 1 \\ k &= 2, 3, \ldots \end{aligned} \tag{3.3}$$

Hence, the sequence reads $F_0 = 1, F_1 = 1, F_2 = 2, F_3 = 3, F_4 = 5, F_5 = 8, F_6 = 13, \ldots$. In the ith iteration step, the ratio for the interval reduction is calculated as

$$\lambda_{(i)} = \frac{F_{N-i}}{F_{N-i+1}}, \; i = 1, 2, \ldots N-2 \tag{3.4}$$

for a precalculated sequence number ending with F_N. Hence, one needs to reach F_N for the number of $N-2$ iteration steps. However, the question is how N is to be selected.

It can be shown that the reduced interval length after $N-2$ iteration steps reads

$$b_{(N-2)} - a_{(N-2)} = \frac{2}{F_N}(b-a) \tag{3.5}$$

Considering (3.5) as the required accuracy ε, the final length must be less or equal to that the selected accuracy, which yields

$$F_N \geq \frac{2}{\varepsilon}(b-a) \tag{3.6}$$

The inner points of a current interval are symmetric concerning the interval bounds. The idea is to "guard" the inner point with the "better" function value.

Algorithm 3.1 (The Fibonacci method)

1. Let $f(x), a, b$ be given. Select $\varepsilon > 0$ and calculate F_N satisfying (3.6) (e.g., rounded to the nearest higher integer). Calculate Fibonacci sequence numbers $\{F_k\}_{k=0}^{N}$ as per (3.3). Initialize the iteration counter $i = 1$, and let $a_{(1)} = a,\; b_{(1)} = b$.
2. The inner interval points $x_{1,(i)}, x_{2,(i)} \in \left[a_{(i)}, b_{(i)}\right]$ are computed as

$$\begin{aligned} x_{1,(i)} &= b_{(i)} - \frac{F_{N-i}}{F_{N-i+1}}\left|b_{(i)} - a_{(i)}\right| \\ x_{2,(i)} &= a_{(i)} + \frac{F_{N-i}}{F_{N-i+1}}\left|b_{(i)} - a_{(i)}\right| \end{aligned} \tag{3.7}$$

   where $x_{1,(i)} < x_{2,(i)}$, and calculate corresponding function values $f\left(x_{1,(i)}\right), f\left(x_{2,(i)}\right)$, see Fig. 3.1.
3. If $f\left(x_{1,(i)}\right) < f\left(x_{2,(i)}\right)$, then

$$a_{(i+1)} = a_{(i)},\; b_{(i+1)} = x_{2,(i)} \tag{3.8a}$$

   else

$$a_{(i+1)} = x_{1,(i)},\; b_{(i+1)} = b_{(i)} \tag{3.8b}$$

4. If $i < N-2$, then $i = i+1$ and go to step 2; else, stop and return

$$x^* = \frac{a_{(N-2)} + b_{(N-2)}}{2} \tag{3.9}$$

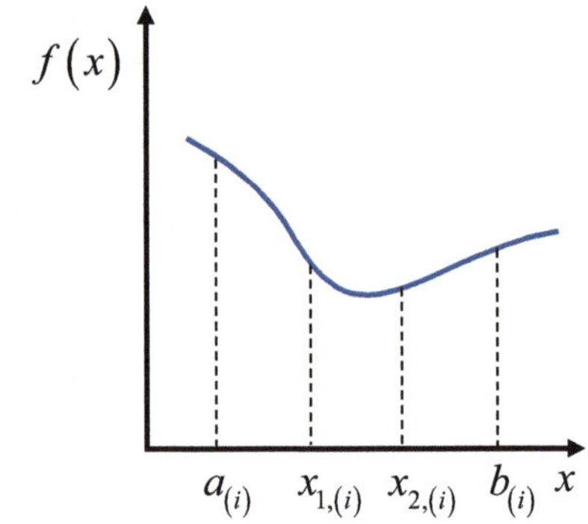

Fig. 3.1 Searching a minimum using Fibonacci and golden ratio methods

Noteworthy, whenever a *maximum* is searched for, the assignment (3.7) is switched.

It is apparent that $f(x)$ must be unimodal on $I = [a, b]$ to receive a global optimum on the interval. Otherwise, the method converges to a local extremum.

The algorithm can be modified and accelerated in such a way that only one inner point is computed in step 2. In each iteration, the current interval already includes one inner point from the preceding iteration. For instance, if a new interval is constructed using (3.8a), the "guarded" point $x_{1,(i)}$ appears as $x_{2,(i+1)}$ in the next iteration. However, new inner points are no longer symmetric with respect to the interval bounds. If one wants to keep these points symmetric, the golden ratio method must be used (see Sect. 3.1.1.2).

A sketch of the Matlab code (for a minimum) follows.

```
% INPUTS
syms f x
f = input('Enter the objective function: ');
e = input('Enter the desired accuracy: ');
interval = input('Enter the given interval [a b]: ');

% INIT
a = interval(1); b = interval(2);
i=1;

% FIBONACCI SEQUENCE
F_N = ceil(2/e*(b-a));
F(1) = 1; F(2) = 1;
k = 2;
while F(k) < F_N
   k = k+1;
   F(k) = F(k-1) + F(k-2);
end


% ITERATIONS
N = length(F);
for k = 1:1:N-3
```

```
    x1 = b - F(N-k)/F(N-k+1)*abs(b-a);
    x2 = a + F(N-k)/F(N-k+1)*abs(b-a);
    if subs(f,x,x1) < subs(f,x,x2)
      b = x2;
    else
      a = x1;
    end
end

% OUTPUT
x_opt = 0.5*(a+b)
```

Example 3.1 *Task*: Find a minimum of $f(x) = -\frac{x}{1+x^2}$ in interval $I = [0, 4]$ using the Fibonacci method with a given accuracy of $\varepsilon = 0.4$.

Solution: First, necessary items of the Fibonacci sequence are given by the value of the F_N calculated using (3.6)

$$F_N \geq \frac{2}{\varepsilon}(b-a) = 5 \cdot 4 = 20$$

which means that

$$F_0 = 1, F_1 = 1, F_2 = 2, F_3 = 3, F_4 = 5, F_5 = 8, F_6 = 13, F_7 = 21$$

It implies that it is necessary to perform $N - 2 - 7 - 2 - 5$ iteration steps. Let $i = 1$, then

$$x_{1,(1)} = b_{(1)} - \frac{F_6}{F_7}\left|b_{(1)} - a_{(1)}\right| = 4 - \frac{13}{21}4 = 1.5238$$

$$x_{2,(1)} - a_{(1)} + \frac{F_6}{F_7}\left|b_{(1)} - a_{(1)}\right| = 0 + \frac{13}{21}4 = 2.4762$$

Objective function values in these inner points are, respectively,

$$f\left(x_{1,(1)}\right) = -0.4587, f\left(x_{2,(1)}\right) = -0.3472$$

As the aim is to find a minimum, the point with the lower function value should be bounded, i.e., $x_{1,(1)}$. Hence, the new interval reads $I_{(2)} = \left[a_{(2)} = a_{(1)}, b_{(2)} = x_{2,(1)}\right] = [0, 2.4762]$.

The iteration counter is incremented to $i = 2$. The inner-interval points are

$$x_{1,(2)} = b_{(2)} - \frac{F_5}{F_6}\left|b_{(2)} - a_{(2)}\right| = 2.4762 - \frac{8}{13}2.4762 = 0.9524$$

$$x_{2,(2)} = a_{(2)} + \frac{F_5}{F_6}\left|b_{(2)} - a_{(2)}\right| = 0 + \frac{8}{13}2.4762 = 1.5238$$

with the corresponding function values

$$f\left(x_{1,(2)}\right) = -0.4994, f\left(x_{2,(2)}\right) = -0.4587$$

The point giving the "better" function value is bounded again, which yields the new interval $I_{(3)} = \left[a_{(3)} = a_{(2)},\ b_{(3)} = x_{2,(2)}\right] = [0,\ 1.5238]$.

The rest of the iteration steps of the algorithm are summarized in Table 3.1.

The eventual interval (after the 5th iteration) is [0.7619, 1.1428], the midpoint of which reads 0.9524. Hence, although the guaranteed interval length is less or equal to 0.4, the actual solution accuracy reaches 0.0486 (i.e., almost an order better).

Figure 3.2 displays the graph of $f(x)$ along with computed interval bounds for each iteration step.

Note that the analytically obtained optimal solution of the problem is $x^* = 1, f^* = -0.5$.

Table 3.1 Fibonacci method—Example 3.1

i	$a_{(i)}$	$b_{(i)}$	$x_{1,(i)}$	$x_{2,(i)}$	$f\left(x_{1,(i)}\right)$	$f\left(x_{2,(i)}\right)$
1	0	4	1.5238	2.4762	– 0.4587	– 0.3472
2	0	2.4762	0.9524	1.5238	– 0.4994	– 0.4587
3	0	1.5238	0.5714	0.9524	– 0.4308	– 0.4994
4	0.5714	1.5238	0.9524	1.1428	– 0.4994	– 0.4956
5	0.5714	1.1428	0.7619	0.9524	– 0.4821	– 0.4994

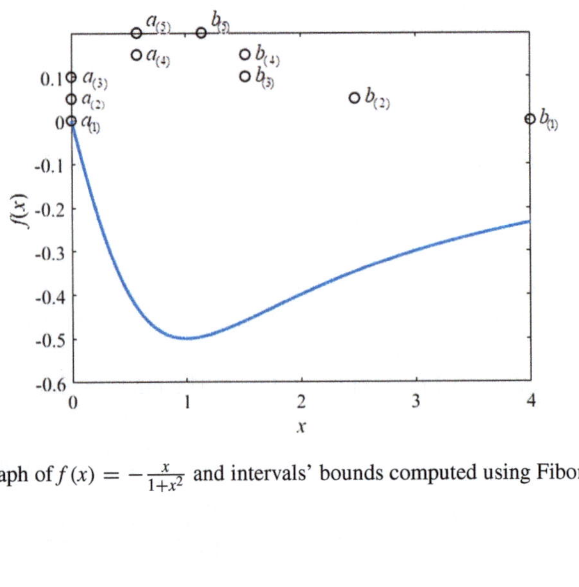

Fig. 3.2 The graph of $f(x) = -\frac{x}{1+x^2}$ and intervals' bounds computed using Fibonacci method—Example 3.1

3.1.1.2 Golden Ratio Method

The golden ratio (or section) is a famous notion known since the ancient age. Although this value (ratio) can be found in nature [15], its introduction can also be motivated by a beauty of technics. For instance, consider a rectangle with side lengths a, b where $a < b$. Now, let a $a \times a$ square is delimited inside the rectangle, see Fig. 3.3. The task is to find ratio $\lambda_g = a/b$ so that the remaining (smaller) rectangle inside the original one has the same ratio of the its sides, i.e., $\lambda_g = (b-a)/a$. It means that BTW, [15]

$$\lambda_g = \frac{a}{b} = \frac{b-a}{a} \tag{3.10}$$

Dividing the right side (3.10) by b, one can obtain the following conditional equality

$$\lambda_g = \frac{1-\lambda_g}{\lambda_g} \Leftrightarrow \lambda_g^2 + \lambda_g - 1 = 0 \tag{3.11}$$

The only positive solution of (3.11) is

$$\lambda_g = \frac{-1+\sqrt{5}}{2} \approx 0.618 \tag{3.12}$$

The golden ratio method is very close to the Fibonacci method. The only difference is that the ratio for the current interval reduction is not variable and calculated using (3.4) (see also step 3 of Algorithm 3.1) but it is constant and given by (3.12), in each iteration step. What is the benefit of such a reduction? The goal is that only one inner point inside the current interval has to be calculated, while the remaining one sits at its place perfectly satisfying the golden ratio (3.13) again. Considering Fig. 3.1, this situation can mathematically be expressed as

$$\lambda = \frac{x_{2,(i)} - a_{(i)}}{b_{(i)} - a_{(i)}} = \frac{b_{(i)} - x_{1,(i)}}{b_{(i)} - a_{(i)}} = \frac{x_{1,(i)} - a_{(i)}}{x_{2,(i)} - a_{(i)}} = \frac{b_{(i)} - x_{2,(i)}}{b_{(i)} - x_{1,(i)}} = \lambda_g \tag{3.13}$$

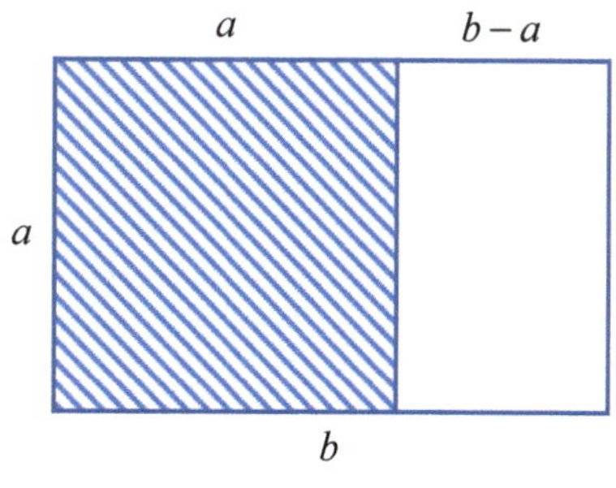

Fig. 3.3 On the notion of the golden ratio

I.e., it can be shown that the ratio from (3.13) must equal those from (3.11). In other words, steps 2 and 3 of Algorithm 3.1 are modified as follows. If $f\left(x_{1,(i)}\right) < f\left(x_{2,(i)}\right)$, only $x_{1,(i+1)}$ is computed in the next iteration as per (3.17) with $\lambda_g \approx 0.618$, while $x_{2,(i+1)} = x_{1,(i)}$. Contrariwise, $x_{2,(i+1)}$ must be recomputed but $x_{1,(i+1)} = x_{2,(i)}$ remains from the preceding iteration step.

The exciting question is also how value (3.12) relates to the Fibonacci sequence. Assume the limit ratio of two consecutive Fibonacci sequence items

$$\lambda = \lim_{i\to\infty} \frac{F_{i-2}}{F_{i-1}} = \lim_{i\to\infty} \frac{F_{i-1}}{F_i} = \lim_{i\to\infty} \frac{F_{i-1}}{F_{i-1} + F_{i-2}} \tag{3.14}$$

Surprisingly, dividing the right-hand expression of (3.14) by F_{i-1}, it is obtained

$$\lambda = \lim_{i\to\infty} \frac{1}{1 + \frac{F_{i-2}}{F_{i-1}}} = \frac{1}{1+\lambda} \tag{3.15}$$

which leads to (3.11) again. Hence, the golden ratio $\lambda_g \approx 0.618$ equals the limit ratio of two consecutive Fibonacci sequence items.

Computation of the necessary number of iteration steps (N) is straightforward. In each step, the interval length is multiplied by λ_g, i.e.,

$$\varepsilon \le (b-a)\lambda_g^N \Rightarrow N \ge \log_{\lambda_g} \frac{\varepsilon}{b-a} = \log_{0.618} \frac{\varepsilon}{b-a} \tag{3.16}$$

The complete algorithm and a Matlab code for searching a minimum follow.

Algorithm 3.2 (The golden ratio method)

1. Let $f(x), a, b$ be given. Select ε and calculate N satisfying (3.16) (rounded to the nearest higher integer). Initialize the iteration counter $i = 1$, and let $a_{(1)} = a,\ b_{(1)} = b$.
2. The initial inner interval points $x_{1,(i)}, x_{2,(i)} \in \left[a_{(i)}, b_{(i)}\right]$ are computed as

$$\begin{aligned} x_{1,(i)} &= b_{(i)} - \lambda_g \left|b_{(i)} - a_{(i)}\right| \\ x_{2,(i)} &= a_{(i)} + \lambda_g \left|b_{(i)} - a_{(i)}\right| \end{aligned} \tag{3.17}$$

   and calculate corresponding function values $f\left(x_{1,(i)}\right), f\left(x_{2,(i)}\right)$.
3. If $f\left(x_{1,(i)}\right) < f\left(x_{2,(i)}\right)$, then

$$\begin{aligned} &a_{(i+1)} = a_{(i)},\ b_{(i+1)} = x_{2,(i)} \\ &x_{1,(i+1)} = b_{(i+1)} - \lambda_g \left|b_{(i+1)} - a_{(i+1)}\right| \\ &x_{2,(i+1)} = x_{1,(i)} \end{aligned} \tag{3.18a}$$

   else

$$a_{(i+1)} = x_{1,(i)},\ b_{(i+1)} = b_{(i)}$$
$$x_{1,(i+1)} = x_{2,(i)} \tag{3.18b}$$
$$x_{2,(i+1)} = a_{(i+1)} + \lambda_g \left| b_{(i+1)} - a_{(i+1)} \right|$$

4. If $i < N$, then $i = i + 1$ and go to step 3; else, stop and return

$$x^* = \frac{a_{(N)} + b_{(N)}}{2} \tag{3.19}$$

```
% INPUTS
syms f x
f = input('Enter the objective function: ');
e = input('Enter the desired accuracy: ');
interval = input('Enter the given interval [a b]: ');

% INIT
a = interval(1); b = interval(2);
i=1;
lam_g=(-1+sqrt(5))/2;

% COMPUTATION OF N
N = ceil(log(e/(b-a))/log(lam_g));

% ITERATIONS
x1 = b - lam_g*abs(b-a);
x2 = a + lam_g*abs(b-a);
for k = 2:1:N
   if subs(f,x,x1) < subs(f,x,x2)
      b = x2;
      x2 = x1;
      x1 = b - lam_g*abs(b-a);
   else
      a = x1;
      x1 = x2;
      x2 = a + lam_g*abs(b-a);
   end
end

% OUTPUT
x_opt = 0.5*(a+b)
```

Example 3.2 *Task*: Find a minimum of $f(x) = -\frac{x}{1+x^2}$ in interval $I = [0, 4]$ using the golden ratio method with a given accuracy of $\varepsilon = 0.4$.

Solution: Formula (3.16) returns $N \geq 4.784$, i.e., perform 5 iteration steps again. Using (3.17), the initial inner points of the original interval are

Table 3.2 Golden ratio method—Example 3.2

i	$a_{(i)}$	$b_{(i)}$	$x_{1,(i)}$	$x_{2,(i)}$	$f(x_{1,(i)})$	$f(x_{2,(i)})$
1	0	4	1.5279	2.4721	– 0.4582	– 0.3476
2	0	2.4721	0.9443	1.5279	– 0.4992	– 0.4582
3	0	1.5279	0.5836	0.9443	– 0.4353	– 0.4992
4	0.5836	1.5279	0.9443	1.1672	– 0.4992	– 0.4941
5	0.5836	1.1672	0.8065	0.9443	– 0.4887	– 0.4992

$$x_{1,(1)} = b_{(1)} - \lambda_g \left| b_{(1)} - a_{(1)} \right| = 4 - 0.618 \cdot 4 = 1.5279$$
$$x_{2,(1)} = a_{(1)} + \lambda_g \left| b_{(1)} - a_{(1)} \right| = 0 + 0.618 \cdot 4 = 2.4721$$

that give the function values, respectively,

$$f(x_{1,(1)}) = -0.4582, \; f(x_{2,(1)}) = -0.3476$$

Since $x_{1,(1)}$ gives "better" function value, the new interval reads $I_{(2)} = \left[a_{(2)} = a_{(1)}, \, b_{(2)} = x_{2,(1)}\right] = [0, \, 2.4721]$ with $x_{2,(2)} = x_{1,(1)}$ being a new right-hand side inner point. It can easily be verified that this inner point does not need to be computed, because it already holds that $x_{2,(2)} = a_{(2)} + \lambda_g \left| b_{(2)} - a_{(2)} \right| = 0 + 0.618 \cdot 2.4721 = 1.5279$ perfectly. However, the left-hand side inner point of the new interval has to be computed as

$$x_{1,(2)} = b_{(2)} - \lambda_g \left| b_{(2)} - a_{(2)} \right| = 2.4721 - 0.618 \cdot 2.4721 = 0.9443$$

Complete results of the example are summarized in Table 3.2.

The eventual interval after the 5th iteration step is [0.8065, 1.1672], the midpoint of which reads 0.9868. I.e., the achieved accuracy is 0.0132. Apparently, results of Examples 3.1 and 3.2 do not significantly differ; however, the golden ratio method works better a bit for the given objective function.

3.1.1.3 Powell (Quadratic Interpolation) Method

Consider the task introduced in Sect. 3.1.1 again, i.e., the seeking of an extremum of single-dimensional unimodular function $f(x)$ in interval $x \in [a, b]$. The Powell (or quadratic interpolation) method belongs to the family of polynomial interpolation methods that are generally based on a polynomial *interpolation* governed by function $\varphi(x)$ of the original objective function $f(x)$. If $\varphi(x)$ is the nth order polynomial, the number of $n + 1$ points and their function values inside the current polynomial must a priori be known. The Powell method uses the simplest meaningful polynomial of the 2nd order, governed by

Fig. 3.4 A principle of the quadratic interpolation

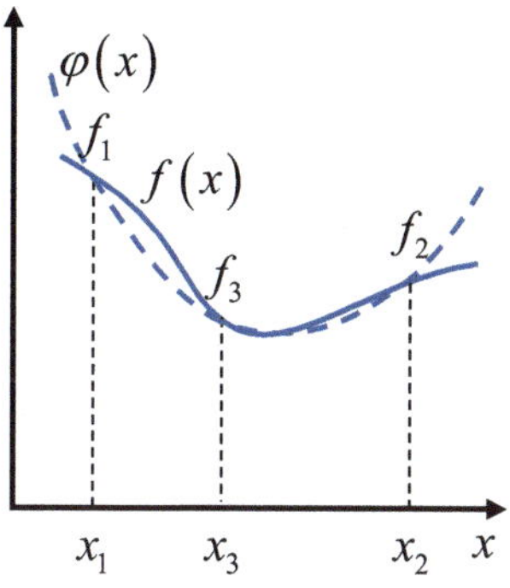

$$\varphi(x) = a_2 x^2 + a_1 x + a_0 \tag{3.20}$$

Let interpolated function vales $f_1 = f(x_1), f_2 = f(x_2), f_3 = f(x_3)$ be known at three points x_1, x_2, x_3 of the current interval, see Fig. 3.4.

From the condition equations $\varphi(x_1) = f(x_1), \varphi(x_2) = f(x_2), \varphi(x_3) = f(x_3)$, the following set of algebraic equations can be determined

$$\begin{aligned} a_2 &= \frac{(f_3 - f_1)(x_2 - x_1) - (f_2 - f_1)(x_3 - x_1)}{\left(x_3^2 - x_1^2\right)(x_2 - x_1) - \left(x_2^2 - x_1^2\right)(x_3 - x_1)} \\ a_1 &= \frac{f_2 - f_1 - a_2\left(x_2^2 - x_1^2\right)}{x_2 - x_1} \\ a_0 &= f_1 - a_1 x_1 - a_2 x_1^2 \end{aligned} \tag{3.21}$$

Then, the extremum point is estimated as the stationary point of $\varphi(x)$ (that is a necessary and sufficient extremum condition for a quadratic function), i.e.,

$$\varphi'(x) = 0 \Leftrightarrow 2a_2 x + a_1 = 0 \Leftrightarrow \tilde{x}^* \approx -\frac{a_1}{2a_2} \tag{3.22}$$

Whenever $a_2 > 0$, $\varphi(x)$ has a local minimum, and vice versa. It means that the type of a local extremum (minimum, maximum) is given by the sign of a_2. Especially, if $f(x)$ is convex on interval $[x_1, x_2]$, the method converges to a local minimum on the interval. If $f(x)$ is concave on the interval, the method converges to a local maximum. If the objective function changes from convex to concave (or vice versa), it might cause a problem, as shown in Example 3.3.

The decision about the interval reduction in the current ith iteration step can adopt two possible principles. First, function values of both inner points $\tilde{x}^*_{(i)}, x_{3,(i)}$ can be evaluated, and the "better" value can be bounded. This strategy, denoted as "Var1" in Algorithm 3.3 below, adopts the idea from both preceding methods. Second, the current extremum estimate $\tilde{x}^*_{(i)}$ is strictly bounded. This strategy is denoted as "Var2". A superiority of "Var1" is demonstrated in Algorithm 3.3 and Example 3.3.

The current interval length can serve as the decision about the iteration termination.

The complete algorithm and a Matlab code for searching a minimum follow. Note that the notation for iteration steps is omitted for simplicity.

Algorithm 3.3 (The Powell method)

1. Let $f(x), a, b$ be given. Select $\varepsilon > 0$ and the interval reduction variant *Var* (optional). Set $x_1 = a,\ x_2 = b, x_3 = 0.5(a + b)$.
2. Evaluate $f_1 = f(x_1), f_2 = f(x_2), f_3 = f(x_3)$.
3. Compute approximation quadratic polynomial ($\varphi(x)$) coefficients using (3.21).
4. Calculate an extremum $\tilde{x}^*$ of $\varphi(x)$ via (3.22) and evaluate $\tilde{f}^* = f(\tilde{x}^*)$.
5.
   (a) If $Var = Var1$:
   If $\tilde{f}^* < f_3$ then do: If $\tilde{x}^* < x_3$ then $x_2 = x_3$; else $x_1 = x_3$. Set $x_3 = \tilde{x}^*$.
   If $\tilde{f}^* \geq f_3$ then do: If $\tilde{x}^* < x_3$ then $x_1 = \tilde{x}^*$; else $x_2 = \tilde{x}^*$.
   (b) If $Var = Var2$:
   If $\tilde{x}^* < x_3$ then $x_2 = x_3$; else $x_1 = x_3$. Set $x_3 = \tilde{x}^*$.
6. If $(x_2 - x_1) \geq \varepsilon$, go to step 2; else, stop and return $x^* = \tilde{x}^*$.

```
% INPUTS
syms f x
f = input('Enter the objective function: ');
e = input('Enter the desired accuracy: ');
interval = input('Enter the given interval [a b]: ');
variant = input('Enter the desired variant (1 = Var1, 2 = Var2): ');

% INIT
x1 = interval(1); x2 = interval(2);
x3 = 0.5*(x1+x2);

% ITERATIONS
while (x2-x1) >= e
   % APPROXIMATING QUADRATIC POLYNOMIAL COEFFS
   f1 = eval(subs(f,x,x1)); f2 = eval(subs(f,x,x2));
   f3 = eval(subs(f,x,x3));
   a2 = ((f3-f1)*(x2-x1)-(f2-f1)*(x3-x1))...
   .../((x3^2-x1^2)*(x2-x1)-(x2^2-x1^2)*(x3-x1));
   a1 = (f2-f1-a2*(x2^2-x1^2))/(x2-x1);
   % a0 = f1 - a1*x1 - a2*x1^2;

   % APPROXIMATING POLYNOMIAL EXTREMUM
   x_opt = -0.5*a1/a2;
   fx_opt = eval(subs(f,x,x_opt));


   % INTERVAL REDUCTION
   if variant == 1
```

```
        if fx_opt < f3
           if x_opt < x3
              x2 = x3;
           else
              x1 = x3;
           end
           x3 = x_opt;
        else
           if x_opt < x3
              x1 = x_opt;
           else
              x2 = x_opt;
           end
        end
     elseif variant == 2
        if x_opt < x3
           x2 = x3;
        else
           x1 = x3;
        end
        x3 = x_opt;
     else
        return
     end
  end

  % OUTPUT
  x_opt
```

Example 3.3 *Task*: Find a minimum of $f(x) = -\frac{x}{1+x^2}$ in interval $I = [0, 4]$ using the Powell method with a given accuracy of $\varepsilon = 0.4$.

Solution: Set $x_1 = 0,\ x_2 = 4, x_3 = 2$. The corresponding function values read

$$f_1 = f(x_1) = 0, f_2 = f(x_2) = -0.2353, f_3 = f(x_3) = -0.4$$

The interpolating quadratic polynomial coefficients are computed via (3.21)

$$a_2 = 0.0706, a_1 = -0.3412, a_0 = 0 \Rightarrow \varphi(x) = 0.0706x^2 - 0.3412x$$

A graphical comparison of $f(x)$ and $\varphi(x)$ is made in Fig. 3.5.
Using (3.22), a minimum of $\varphi(x)$ reads

$$\tilde{x}^* = -\frac{a_1}{2a_2} = 2.4167$$

with the corresponding function value of $f(\tilde{x}^*) = -0.3533$.

Let Var1 be considered. Then, according to step 5 of Algorithm 3.3, $x_3 = 2$ must remain bounded in the new (reduced) interval $I = [0,\ 2.4167]$. Now, one has

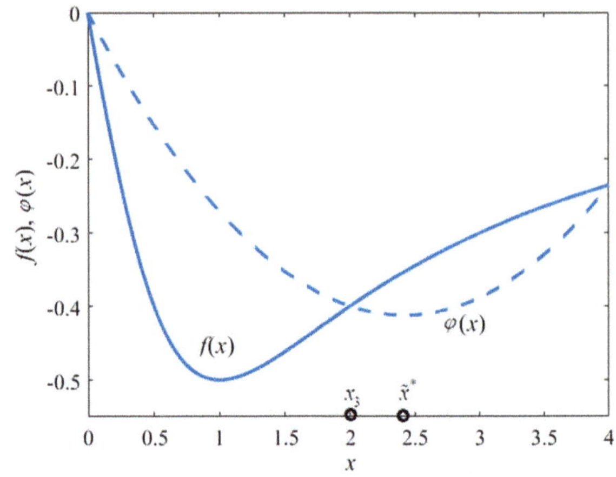

Fig. 3.5 Graphs of $f(x) = -\frac{x}{1+x^2}$ and $\varphi(x) = 0.0706x^2 - 0.3412x$ with the interval midpoint x_3 and an extremum $\tilde{x}^*$ of $\varphi(x)$—Example 3.3

$x_1 = 0$, $x_2 = 2.4167, x_3 = 2$, giving rise to $f_1 = 0, f_2 = -0.3533, f_3 = -0.4$, respectively. The interpolating polynomial reads

$$\varphi(x) = 0.1291x^2 - 0.4583x$$

with its finite extremum $\varphi(\tilde{x}^*) = -0.4277$ at $\tilde{x}^* = 1.7742$. The decision rule implies the reduced interval $I = [0,\ 2]$ with the inner point $x_3 = 1.7742$.

The remaining data for Var1 and the five iteration steps is summarized in Table 3.3.

Notice that the convergence rate is too low. The required accuracy of $\varepsilon = 0.4$ is suddenly reached after 48 iterations; however, the estimated extremum is almost perfect.

It is clear from Fig. 3.5 that the Powell method using Var2 can fail for non-convex functions. The problem is that a minimum of $\varphi(x)$ lies in another half-interval than the actual minimum of $f(x)$ is. Indeed, according to Var2, the initial interval reduces to $I = [2,\ 4]$. It means that further interval reductions cannot reach the extremum at $x^* = 1$, as it lies outside this interval.

Besides using Var1, another possibility can be using a more complex (e.g., cubic) interpolating polynomial; however, it is computationally more demanding.

Table 3.3 Powell method (Var1)—Example 3.3

i	$a = x_1$	$b = x_2$	x_3	$\tilde{x}^*$	$f(x_3)$	$f(\tilde{x}^*)$
1	0	4	2	2.4167	− 0.4	− 0.3533
2	0	2.4167	2	1.7742	− 0.4	− 0.4277
3	0	2	1.7742	1.5495	− 0.4277	− 0.4556
4	0	1.7742	1.5495	1.3987	− 0.4556	− 0.4731
5	0	1.5495	1.3987	1.2761	− 0.4731	− 0.4855

3.1.2 Multi-dimensional Unconstrained Problem

Consider the unconstrained multi-dimensional optimization problem defined in (2.15). The following two methods presented in this section search the variable space using the so-called simplex that represents a specific geometric notion.

Definition 3.2 A **simplex** in the $\mathbb{R}^n$ space is the n-dimensional polytope which is the convex hull with $n + 1$ vertices.

To rephrase Definition 3.2, a simplex represents "the smallest" possible convex figure in the given dimension. For instance, it is a triangle in $\mathbb{R}^2$, a tetrahedron in $\mathbb{R}^3$, etc. A simplex can be *equilateral* or *regular* (i.e., all its edges have the same length), or irregular.

The third method of this section inspects the variable space via orthogonal jumps with a single-dimensional suboptimization.

3.1.2.1 Regular Simplex Method

This method is also called (after its inventors) the **Spendley–Hext–Himsworth** method [21]. The method is based on searching the variable space using "flipping" regular simplexes. Always, the vertex with the "worst" functional value is flipped symmetrically over the "center" of all other vertices. Let all the steps be described before the complete algorithm is presented.

First, a regular simplex must be constructed based on the selected initial point $\mathbf{x}_{(1)}^1$ and the simplex edge length Δ, for a given objective function $f(\mathbf{x})$. The regular simplex vertices are generated via the following rule

$$\mathbf{x}_{(1)}^j = \mathbf{x}_{(1)}^1 + d\mathbf{1} + l\mathbf{e}_{j-1},\ j = 2, 3, \ldots n + 1 \tag{3.23}$$

where

$$\begin{aligned} \mathbf{1} &= \begin{bmatrix} 1\ 1 \ldots 1 \end{bmatrix}^T \in \mathbb{R}^n \\ d &= \Delta\frac{\sqrt{n+1} - 1}{n\sqrt{2}},\ l = \frac{\Delta}{\sqrt{2}} \end{aligned} \tag{3.24}$$

and $\mathbf{e}_j$ means the unit Euclidean vector of dimension n with 1 at jth position.

Each vertex represents an agent that inspects the particular functional value. Hence, evaluate $f\left(\mathbf{x}_{(1)}^j\right)$, $j = 1, 2, \ldots n + 1$. If the task is to minimize $f(\mathbf{x})$, the leading idea is to flip the "worst" vertex (i.e., that with the highest function value)

$$\mathbf{x}_{(i)}^h = \arg\max f\left(\mathbf{x}_{(i)}^j\right),\ j = 1, 2, \ldots n + 1 \tag{3.25}$$

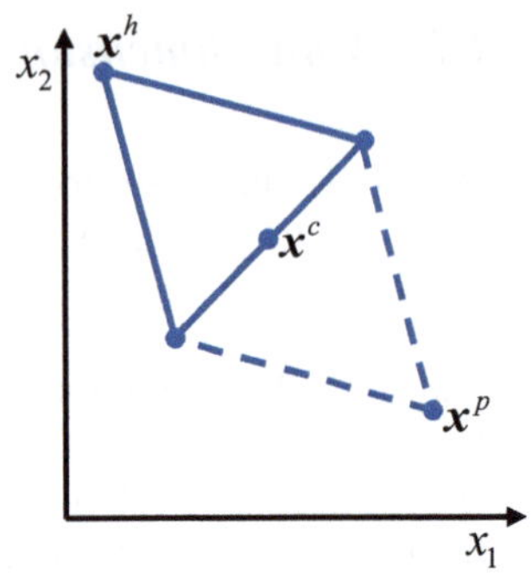

Fig. 3.6 On the flipping of $\mathbf{x}^h$—the regular simplex method

over the center $\mathbf{x}^c_{(i)}$ of other (remaining) vertices in each iteration step i. This center can simply be computed as the arithmetical mean of the remaining vertices

$$\mathbf{x}^c_{(i)} = \frac{1}{n}\sum_{\substack{j=1 \\ j\neq h}}^{n+1} \mathbf{x}^j_{(i)} \tag{3.26}$$

The flipping vector operation reads

$$\mathbf{x}^p_{(i)} = \mathbf{x}^c_{(i)} + \left(\mathbf{x}^c_{(i)} - \mathbf{x}^h_{(i)}\right) = 2\mathbf{x}^c_{(i)} - \mathbf{x}^h_{(i)} \tag{3.27}$$

Then, $\mathbf{x}^h_{(i)}$ is canceled in the following step and $\mathbf{x}^p_{(i)}$ represents a new vertex of the simplex, see Fig. 3.6.

However, this scenario does not always apply. The issue is that situations where some kind of looping of the algorithm may occur must be detected and resolved. In principle, two possible looping schemas can be identified.

First, what if the flipped vertex is the worst one in the next iteration step, i.e., $\mathbf{x}^p_{(i)} = \mathbf{x}^h_{(i+1)}$? It is clear from Fig. 3.6 that it will hold $\mathbf{x}^p_{(i+1)} = \mathbf{x}^h_{(i)}$ without any special action and, hence, the algorithm loops. In this case, the "second worst" vertex

$$\mathbf{x}^s_{(i)} = \arg\max f\left(\mathbf{x}^j_{(i)}\right),\ j = 1, 2, \ldots n+1,\ j \neq h \tag{3.28}$$

is flipped instead of $\mathbf{x}^h_{(i)}$.

Second, the simplex symmetry may cause the flipped vertex positions to repeat in the variable space. Such a scenario may occur if any of the vertices (say $\mathbf{x}^q$) is not flipped for a certain number of iterations (two in $\mathbb{R}$, five in $\mathbb{R}^2$, minimally eight in $\mathbb{R}^3$); see Fig. 3.7.

The recommended solution is to *reduce* the simple so that its edge length is shortened and all the vertices are moved towards $\mathbf{x}^q$

$$\mathbf{x}^j_{(i+1)} = \mathbf{x}^j_{(i+1)} + \alpha\left(\mathbf{x}^q_{(i+1)} - \mathbf{x}^j_{(i+1)}\right)$$

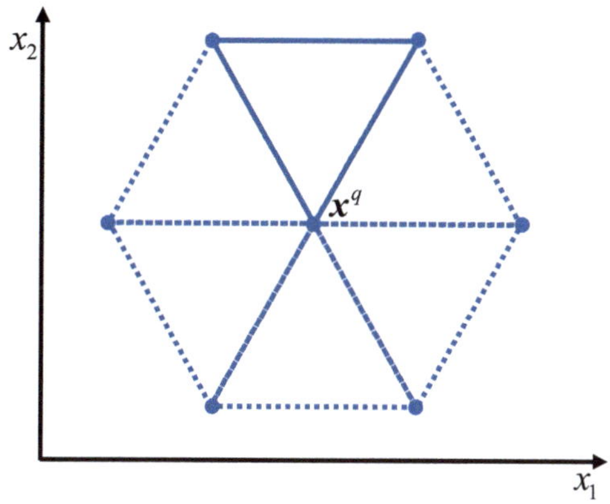

Fig. 3.7 On a possible looping of the regular simplex method

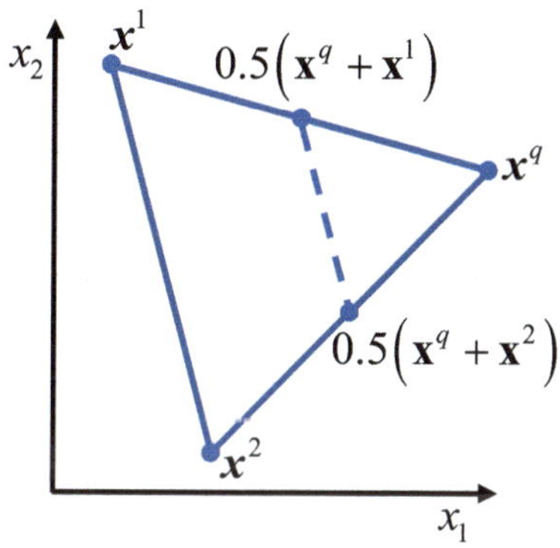

Fig. 3.8 Regular simplex reduction (3.29) with $\alpha = 0.5$

$$
\begin{aligned}
&= \alpha \mathbf{x}^q_{(i+1)} + (1-\alpha)\mathbf{x}^j_{(i+1)},\ j = 1, 2, \ldots, n+1,\ j \neq q \\
\Delta &= \alpha \Delta,\ \alpha \in (0, 1)
\end{aligned}
\tag{3.29}
$$

Figure 3.8 illustrates this shifting for $\alpha = 0.5$.

A complete algorithm of the method when searching for a minimum of $f(\mathbf{x})$ follows.

Algorithm 3.4 (The regular simplex method)

1. Let $f(\mathbf{x})$ be given. Select the initial point $\mathbf{x}^1_{(1)}$, the initial simplex edge length Δ, the terminating simplex edge length Δ_ε and the reduction factor $\alpha \in (0, 1)$. Initialize the iteration steps counter $i = 1$.
2. Construct the initial regular simple using (3.23) and (3.24). Initialize indicator vector $\boldsymbol{\Omega} = [0, 0, \ldots, 0] \in \mathbb{Z}^{n+1}$ to check a special kind of looping.
3. Evaluate function values at vertices $f\left(\mathbf{x}^j_{(i)}\right)$, $j = 1, 2, \ldots n+1$ and find $\mathbf{x}^h_{(i)}$ as per (3.25).

4. Compute center $\mathbf{x}^c_{(i)}$ using (3.26) and perform flipping (3.27) to get the flipped point $\mathbf{x}^p_{(i)}$.
5. If $\mathbf{x}^h_{(i)} = \mathbf{x}^p_{(i-1)}$ for $i > 1$, find $\mathbf{x}^s_{(i)}$ via (3.28) and formally set $\mathbf{x}^h_{(i)} = \mathbf{x}^s_{(i)}$.
6. Set $\mathbf{\Omega}(j) = \mathbf{\Omega}(j) + 1, j = 1, 2, \ldots, n+1, j \neq h$, $\mathbf{\Omega}(h) = 0$.
7. If $\exists j : \mathbf{\Omega}(j) \geq 3n + 1$ then set $\mathbf{\Omega} = [0, 0, \ldots, 0]$ and go to step 8; otherwise, go to step 8.
8. Perform the reduction as per (3.29), set $\Delta = (1 - \alpha)\Delta$, and go to step 10.
9. Perform the flipping as per (3.26) and (3.27), set $\mathbf{x}^h_{(i+1)} = \mathbf{x}^p_{(i)}$ (with respect to vertices indexing), $i = i + 1$, and go to step 3.
10. If $\Delta \geq \Delta_\varepsilon$, set $i = i + 1$ and go to step 3; else, stop and return

$$\mathbf{x}^* = \arg\min f\left(\mathbf{x}^j_{(i)}\right), \ j = 1, 2, \ldots n + 1 \tag{3.30}$$

A sketch of the Matlab code for a two-variable function (for simplicity) follows.

```
% INPUTS
syms f x1 x2
f = input('Enter the objective function: ');
x_init = input('Enter the initial point [x1 x2]: ');
Delta = input('Enter the initial simplex edge length: ');
Delta_e = input('Enter the desired accuracy (final simplex edge
length): ');
alpha = input('Enter the simplex reduction factor (alpha > 0): ');

% INIT
n = 2;
i = 1;
X(1,1:n) = x_init;
F(1) = eval(subs(f,[x1 x2],X(1,:)));

I = eye(n);
d = Delta*((sqrt(n+1)-1)/(n*sqrt(2)));
l = Delta/sqrt(2);
for j = 2:1:(n+1)
   X(j,1:n) = x_init + d*ones(1,n) + l*I(j-1,:);
   F(j) = eval(subs(f,[x1 x2],X(j,:)));
end

Sigma = zeros(1,n+1);

% ITERATIONS
while Delta >= Delta_e
   % SEARCHING FOR THE WORST VERTEX
   [F_sort, i_sort] = sort(F);
   ih = i_sort(end);
   xh = X(ih,1:n);


   % THE FIRST KIND OF LOOPING
```

```
    if i>1
       if ih==ip
          ih = i_sort(end-1);
          xh = X(ih,1:n);
       end
    end

    Sigma = Sigma + ones(1,n+1); Sigma(ih) = 0;

    % THE SECOND KIND OF LOOPING - REDUCTION
    [S_m, iq] = max(Sigma);
    if S_m >= (3*n+1)
       for j = 1:1:(n+1)
          X(j,1:n) = alpha*X(j,1:n) + (1-alpha)*X(iq,1:n);
          F(j) = eval(subs(f,[x1 x2],X(j,:)));
       end
       Sigma = zeros(1,n+1);
       Delta = alpha*Delta;
       ip=0;
    else
       % FLIPPING
       if ih==1
          xc = sum(X(2:end,1:n))/n;
       elseif ih==(n+1)
          xc = sum(X(1:ih-1,1:n))/n;
       else
          xc = sum(X([1:ih-1,ih+1:end],1:n))/n;
       end
       xp = 2*xc - xh;
       X(ih,1:n)=xp;
       F(ih) = eval(subs(f,[x1 x2],xp));
       ip = ih;
    end
    i=i+1;
end

% OUTPUT
X(i_sort(1),1:n)
```

Example 3.4 *Task*: Find a minimum of $f(\mathbf{x}) = f(x_1, x_2) = 3x_1^2 - 2x_1x_2 + x_2^2 + 4x_1 + 3x_2 + 1$ on $\mathbb{R}^2$ using the regular simplex (Spendley–Hext–Himsworth) method. Let the initial point be $\mathbf{x}_{(1)}^1 = [0, 0]^T$, the initial simplex edge length $\Delta = 1$, and the reduction factor $\alpha = 0.5$. Choose the fixed number of $N = 15$ iteration steps. What is the eventual simplex edge length?

Solution: Let the first two iterations steps be shown in detail. The initial simplex is constructed using (3.23) and (3.24), giving rise to

$$d = \Delta\frac{\sqrt{n+1}-1}{n\sqrt{2}} = \frac{\sqrt{2+1}-1}{2\sqrt{2}} = 0.2588,\ l = \frac{\Delta}{\sqrt{2}} = \frac{1}{\sqrt{2}} = 0.7071$$

$$\begin{aligned}
\mathbf{x}_{(1)}^1 &= [0, 0]^T \\
\mathbf{x}_{(1)}^2 &= \mathbf{x}_{(1)}^1 + d\mathbf{1} + l\mathbf{e}_1 = [0, 0]^T + 0.2588 \cdot [1, 1]^T + 0.7071 \cdot [1, 0]^T \\
&= [0.9659, 0.2588]^T \\
\mathbf{x}_{(1)}^3 &= \mathbf{x}_{(1)}^1 + d\mathbf{1} + l\mathbf{e}_2 = [0, 0]^T + 0.2588 \cdot [1, 1]^T + 0.7071 \cdot [0, 1]^T \\
&= [0.2588, 0.9659]^T
\end{aligned}$$

The corresponding function values read $f\left(\mathbf{x}_{(1)}^1\right) = 1, f\left(\mathbf{x}_{(1)}^2\right) = 8.0062$, $f\left(\mathbf{x}_{(1)}^3\right) = 5.567$. The highest (i.e., the worst) function value is at vertex $\mathbf{x}_{(1)}^h = \mathbf{x}_{(1)}^2$ that ought to be flipped. Compute the center point and the flipped point

$$\mathbf{x}_{(1)}^c = \frac{\mathbf{x}_{(1)}^1 + \mathbf{x}_{(1)}^3}{2} = [0.1294, 0.4829]^T,\ \mathbf{x}_{(1)}^p = 2\mathbf{x}_{(1)}^c - \mathbf{x}_{(1)}^h = [-0.7071, 0.7071]^T$$

Notice that $f\left(\mathbf{x}_{(1)}^p\right) = 3.2929$ is "better" than $f\left(\mathbf{x}_{(1)}^h\right)$, i.e., $f\left(\mathbf{x}_{(1)}^p\right) < f\left(\mathbf{x}_{(1)}^h\right)$. The new simplex is $\mathbf{x}_{(2)}^1 = \mathbf{x}_{(1)}^1$, $\mathbf{x}_{(2)}^2 = \mathbf{x}_{(1)}^p$, $\mathbf{x}_{(2)}^3 = \mathbf{x}_{(1)}^3$.

Hence, one has $f\left(\mathbf{x}_{(2)}^1\right) = 1, f\left(\mathbf{x}_{(2)}^2\right) = 3.2929, f\left(\mathbf{x}_{(2)}^3\right) = 5.567$ for $i = 2$. It can be computed that $\mathbf{x}_{(2)}^c = [-0.3535, 0.3535]^T$, $\mathbf{x}_{(2)}^p = [-0.9659, -0.2588]^T$, and $f\left(\mathbf{x}_{(2)}^p\right) = -1.2741$. The vertex $\mathbf{x}_{(2)}^p$ substitutes $\mathbf{x}_{(2)}^3$ for the next iteration step.

The remaining iterations are summarized in Table 3.4.

Some notes to Table 3.4:

- For $i = 7$ and $i = 14$, the objective function f has a higher value at the flipped point $\mathbf{x}_{(i)}^p$ than at the flipping one $\mathbf{x}_{(i)}^h$. Hence, $\mathbf{x}_{(i)}^s$ is flipped in the next iteration step.
- For $i = 11$, the point $\mathbf{x}_{(i)}^s$ has not been flipped for five preceding steps; therefore, the simplex is reduced, giving rise to a new one formulated in $i = 12$.
- The simplex edge length after the 15th iteration is $\Delta = 0.5$.

The eventual simplex is composed of vertices $\mathbf{x}^1 = [-1.4836, -2.1907]^T$, $\mathbf{x}^2 = [-1.613, -2.6736]^T$, and $\mathbf{x}^3 = [-2.0959, -2.803]^T$. Either the vertex with the lowest function value (i.e., $f(\mathbf{x}^2) = -7.144$) can be taken as the minimum estimation or the simplex center can be calculated: $\mathbf{x}^C = \sum_{i=1}^3 \mathbf{x}^i/3 = [-1.7308, -2.5558]^T$, $f(\mathbf{x}^C) = -6.9186$.

The analytic local and global minimum lies at $\mathbf{x}^* = [-1.75, -3.25]^T$ with $f(\mathbf{x}^*) = -7.375$, see Sect. 2.2.

The evolution of the position and size of the regular simple according to Example 3.4 (Table 3.4) along with the shape of $f(x_1, x_2)$ are displayed in Fig. 3.9.

Table 3.4 Regular simplex method—Example 3.4

i	Δ	$\mathbf{x}^1_{(i)}$	$\mathbf{x}^2_{(i)}$	$\mathbf{x}^3_{(i)}$	$\mathbf{x}^p_{(i)}$	$f\left(\mathbf{x}^1_{(i)}\right)$	$f\left(\mathbf{x}^2_{(i)}\right)$	$f\left(\mathbf{x}^3_{(i)}\right)$	$f\left(\mathbf{x}^p_{(i)}\right)$
1	1	$\begin{bmatrix} 0 \\ 0 \end{bmatrix}$	$\begin{bmatrix} 0.9659 \\ 0.2588 \end{bmatrix}$	$\begin{bmatrix} 0.2588 \\ 0.9659 \end{bmatrix}$	$\begin{bmatrix} -0.7071 \\ 0.7071 \end{bmatrix}$	1	− 1.0062	5.567	3.2929
2	1	$\begin{bmatrix} 0 \\ 0 \end{bmatrix}$	$\begin{bmatrix} -0.7071 \\ 0.7071 \end{bmatrix}$	$\begin{bmatrix} 0.2588 \\ 0.9659 \end{bmatrix}$	$\begin{bmatrix} -0.9659 \\ -0.2588 \end{bmatrix}$	1	3.2929	5.567	− 1.274
3	1	$\begin{bmatrix} 0 \\ 0 \end{bmatrix}$	$\begin{bmatrix} -0.7071 \\ 0.7071 \end{bmatrix}$	$\begin{bmatrix} -0.9659 \\ -0.2588 \end{bmatrix}$	$\begin{bmatrix} -0.9659 \\ -0.2588 \end{bmatrix}$	1	3.2929	− 1.2741	− 2.299
4	1	$\begin{bmatrix} 0 \\ 0 \end{bmatrix}$	$\begin{bmatrix} -0.2588 \\ -0.9659 \end{bmatrix}$	$\begin{bmatrix} -0.9659 \\ -0.2588 \end{bmatrix}$	$\begin{bmatrix} -1.2247 \\ -1.2247 \end{bmatrix}$	1	− 2.2991	− 1.2741	− 4.573
5	1	$\begin{bmatrix} -1.2247 \\ -1.2247 \end{bmatrix}$	$\begin{bmatrix} -0.2588 \\ -0.9659 \end{bmatrix}$	$\begin{bmatrix} -0.9659 \\ -0.2588 \end{bmatrix}$	$\begin{bmatrix} -0.5176 \\ -1.9319 \end{bmatrix}$	− 4.5732	− 2.2991	− 1.2741	− 4.330
6	1	$\begin{bmatrix} -1.2247 \\ -1.2247 \end{bmatrix}$	$\begin{bmatrix} -0.2588 \\ -0.9659 \end{bmatrix}$	$\begin{bmatrix} -0.5176 \\ -1.9319 \end{bmatrix}$	$\begin{bmatrix} -1.4836 \\ -2.1907 \end{bmatrix}$	− 4.5732	− 2.2991	− 4.3302	− 6.604
7	1	$\begin{bmatrix} -1.2247 \\ -1.2247 \end{bmatrix}$	$\begin{bmatrix} -1.4836 \\ -2.1907 \end{bmatrix}$	$\begin{bmatrix} -0.5176 \\ -1.9319 \end{bmatrix}$	$\begin{bmatrix} -2.1907 \\ -1.4836 \end{bmatrix}$	− 4.5732	− 6.6043	− 4.3302	− 2.115
8	1	$\begin{bmatrix} -1.2247 \\ -1.2247 \end{bmatrix}$	$\begin{bmatrix} -1.4836 \\ -2.1907 \end{bmatrix}$	$\begin{bmatrix} -2.1907 \\ -1.4836 \end{bmatrix}$	$\begin{bmatrix} -2.4495 \\ -2.4495 \end{bmatrix}$	− 4.5732	− 6.6043	− 2.1153	− 4.146
9	1	$\begin{bmatrix} -2.4495 \\ -2.4495 \end{bmatrix}$	$\begin{bmatrix} -1.4836 \\ -2.1907 \end{bmatrix}$	$\begin{bmatrix} -2.1907 \\ -1.4836 \end{bmatrix}$	$\begin{bmatrix} -1.7424 \\ -3.1566 \end{bmatrix}$	− 4.146	− 6.6043	− 2.1153	− 7.368

(continued)

Table 3.4 (continued)

i	Δ	$\mathbf{x}^1_{(i)}$	$\mathbf{x}^2_{(i)}$	$\mathbf{x}^3_{(i)}$	$\mathbf{x}^p_{(i)}$	$f\left(\mathbf{x}^1_{(i)}\right)$	$f\left(\mathbf{x}^2_{(i)}\right)$	$f\left(\mathbf{x}^3_{(i)}\right)$	$f\left(\mathbf{x}^p_{(i)}\right)$
10	1	$\begin{bmatrix} -2.4495 \\ -2.4495 \end{bmatrix}$	$\begin{bmatrix} -1.4836 \\ -2.1907 \end{bmatrix}$	$\begin{bmatrix} -1.7424 \\ -3.1566 \end{bmatrix}$	$\begin{bmatrix} -0.7765 \\ -2.8978 \end{bmatrix}$	− 4.146	− 6.6043	− 7.3675	− 5.093
11	1	$\begin{bmatrix} -0.7765 \\ -2.8978 \end{bmatrix}$	$\begin{bmatrix} -1.4836 \\ -2.1907 \end{bmatrix}$	$\begin{bmatrix} -1.7424 \\ -3.1566 \end{bmatrix}$	Reduction	− 5.0934	− 6.6043	− 7.3675	Reduct.
12	0.5	$\begin{bmatrix} -1.4836 \\ -2.1907 \end{bmatrix}$	$\begin{bmatrix} -1.0006 \\ -2.0613 \end{bmatrix}$	$\begin{bmatrix} -1.3542 \\ -1.7077 \end{bmatrix}$	$\begin{bmatrix} -1.13 \\ -2.5442 \end{bmatrix}$	− 6.6043	− 6.0588	− 5.7473	− 6.599
13	0.5	$\begin{bmatrix} -1.4836 \\ -2.1907 \end{bmatrix}$	$\begin{bmatrix} -1.0006 \\ -2.0613 \end{bmatrix}$	$\begin{bmatrix} -1.13 \\ -2.5442 \end{bmatrix}$	$\begin{bmatrix} -1.613 \\ -2.6736 \end{bmatrix}$	− 6.6043	− 6.0588	− 6.5989	− 7.144
14	0.5	$\begin{bmatrix} -1.4836 \\ -2.1907 \end{bmatrix}$	$\begin{bmatrix} -1.613 \\ -2.6736 \end{bmatrix}$	$\begin{bmatrix} -1.13 \\ -2.5442 \end{bmatrix}$	$\begin{bmatrix} -1.9665 \\ -2.3201 \end{bmatrix}$	− 6.6043	− 7.144	− 6.5989	− 5.967
15	0.5	$\begin{bmatrix} -1.4836 \\ -2.1907 \end{bmatrix}$	$\begin{bmatrix} -1.613 \\ -2.6736 \end{bmatrix}$	$\begin{bmatrix} -1.9665 \\ -2.3201 \end{bmatrix}$	$\begin{bmatrix} -2.0959 \\ -2.803 \end{bmatrix}$	− 6.6043	− 7.144	− 5.9669	− 6.507

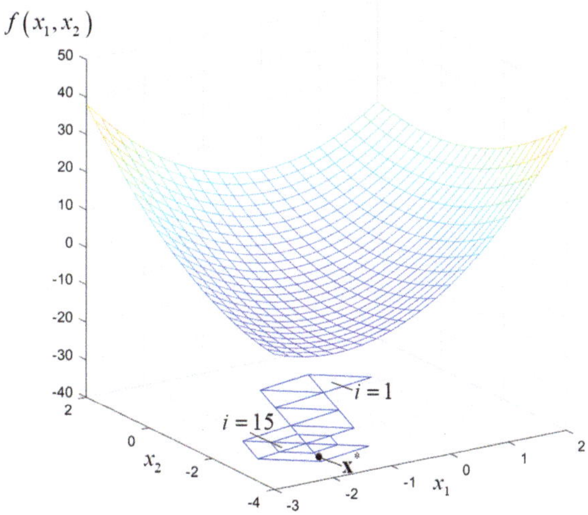

Fig. 3.9 The graph of $f(x_1, x_2) = 3x_1^2 - 2x_1x_2 + x_2^2 + 4x_1 + 3x_2 + 1$ and the regular simplex evolution—Example 3.4

3.1.2.2 Variable Simplex Method

The method is considered the most perfect of the "classical" comparative iteration methods. It was investigated by Nelder and Mead in 1965; therefore, it is also called the **Nelder–Mead method** [16]. The principle of the method is based on a smart deformation of the initially regular or rectangular simplex to reach an extreme faster. At each step, the information about the simplex vertices having the "best", the "second worst", and the "worst" values of the objective function is maintained.

In this section, an explanation of the different simplex deformation strategies is presented, followed by the algorithm of the method for finding a minimum, a Matlab code, and an illustrative example.

The initial simplex may not be regular. Although some recent research has been made on a suitable choice of its vertices, it is possible to create a right (rectangular) simplex as follows:

$$\mathbf{x}_{(1)}^j = \mathbf{x}_{(1)}^1 + \Delta \mathbf{e}_{j-1},\ j = 2, 3, \ldots n + 1 \tag{3.31}$$

where $\mathbf{x}_{(1)}^1$ is the initially selected vertex, Δ means the cathetus length (see Fig. 3.10), and $\mathbf{e}_j$ expresses the unit Euclidean vector of dimension n with 1 at jth position again.

Considering the *minimization* problem, objective function values at each vertex are computed, and three vertices (namely, the "best", "second worst", and "worst") are labeled as

Fig. 3.10 A right (rectangular) simplex with the cathetus length of Δ

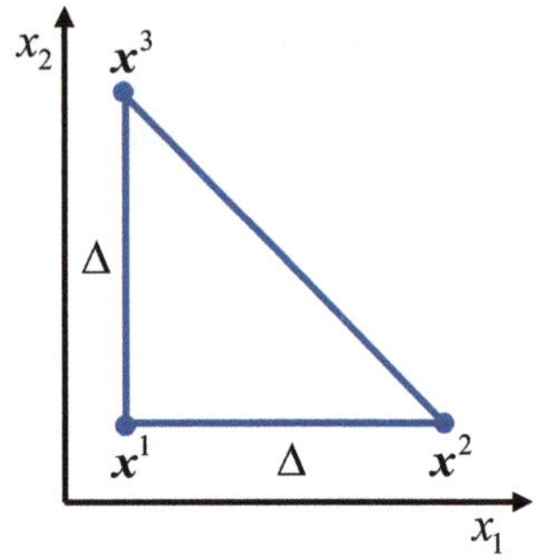

$$\begin{aligned}\mathbf{x}_{(i)}^{h} &= \arg\max f\left(\mathbf{x}_{(i)}^{j}\right), \; j = 1, 2, \ldots n + 1\\ \mathbf{x}_{(i)}^{s} &= \arg\max f\left(\mathbf{x}_{(i)}^{j}\right), \; j = 1, 2, \ldots n + 1, \; j \neq h\\ \mathbf{x}_{(i)}^{l} &= \arg\min f\left(\mathbf{x}_{(i)}^{j}\right), \; j = 1, 2, \ldots n + 1\end{aligned} \tag{3.32}$$

Note that this sorting is made in every iteration step. Then, $\mathbf{x}_{(i)}^{h}$ is flipped over the center $\mathbf{x}_{(i)}^{c}$ of the remaining vertices, analogously to the regular simplex method. The point $\mathbf{x}_{(i)}^{c}$ is computed via (3.26) again.

In contrast to the regular simplex method (see Sect. 3.1.2.1), this method is not satisfied with the simple flipping (also called the **reflection**) but further evaluates the flipped point $\mathbf{x}_{(i)}^{p}$. A particular scenario is applied based on this evaluation.

First, if $f\left(\mathbf{x}_{(i)}^{l}\right) \leq f\left(\mathbf{x}_{(i)}^{p}\right) \leq f\left(\mathbf{x}_{(i)}^{s}\right)$, the current iteration step is finished. It means that $\mathbf{x}_{(i)}^{p}$ becomes a new simplex vertex, and $\mathbf{x}_{(i)}^{h}$ is cancelled. With respect to the indexing of vertices, it can formally be expressed as $\mathbf{x}_{(i+1)}^{h} = \mathbf{x}_{(i)}^{p}$. (Note that the worst vertex for the next iteration is not known at this step, in fact.)

Second, if $f\left(\mathbf{x}_{(i)}^{p}\right) < f\left(\mathbf{x}_{(i)}^{l}\right)$, the idea is to "look behind" $\mathbf{x}_{(i)}^{p}$ since this direction (i.e., $\mathbf{x}_{(i)}^{p} - \mathbf{x}_{(i)}^{c}$) seems to be perspective. Therefore, the so-called **expansion** is performed:

$$\begin{aligned}&\mathbf{x}_{(i)}^{p0} = \mathbf{x}_{(i)}^{c} + \alpha\left(\mathbf{x}_{(i)}^{p} - \mathbf{x}_{(i)}^{c}\right) = \alpha\mathbf{x}_{(i)}^{p} + (1 - \alpha)\mathbf{x}_{(i)}^{c}\\ &\alpha > 1\end{aligned} \tag{3.33}$$

The originally proposed setting is $\alpha = 2$; however, it may be altered by the user (e.g., based on the result of a metaoptimization). The function value of the expanded point $f\left(\mathbf{x}_{(i)}^{p0}\right)$ is further evaluated:

- If $f\left(\mathbf{x}_{(i)}^{p0}\right) < f\left(\mathbf{x}_{(i)}^{l}\right)$, then $\mathbf{x}_{(i)}^{h}$ is substituted by $\mathbf{x}_{(i)}^{p0}$.
- If $f\left(\mathbf{x}_{(i)}^{p0}\right) \geq f\left(\mathbf{x}_{(i)}^{l}\right)$, then $\mathbf{x}_{(i)}^{h}$ is substituted by $\mathbf{x}_{(i)}^{p}$.

The expansion is elucidated in Fig. 3.11. Intuitively, one may consider comparing $\mathbf{x}_{(i)}^{p0}$ with $\mathbf{x}_{(i)}^{p}$; however, the above-given strategy represents the original concept.

Third, whenever $f\left(\mathbf{x}_{(i)}^{s}\right) < f\left(\mathbf{x}_{(i)}^{p}\right) < f\left(\mathbf{x}_{(i)}^{h}\right)$, the idea is to "look back a bit" in the reverse direction to the flipping (i.e., $\mathbf{x}_{(i)}^{c} - \mathbf{x}_{(i)}^{p}$), but not inside the current simplex shape. This operation is called the **outer contraction**, and it is formulated as

$$\begin{aligned} &\mathbf{x}_{(i)}^{p0} = \mathbf{x}_{(i)}^{c} + \beta\left(\mathbf{x}_{(i)}^{p} - \mathbf{x}_{(i)}^{c}\right) = \beta\mathbf{x}_{(i)}^{p} + (1-\beta)\mathbf{x}_{(i)}^{c} \\ &\beta \in (0, 1) \end{aligned} \tag{3.34}$$

see Fig. 3.12.

Note that default setting is $\beta = 0.5$. The additional test follows:

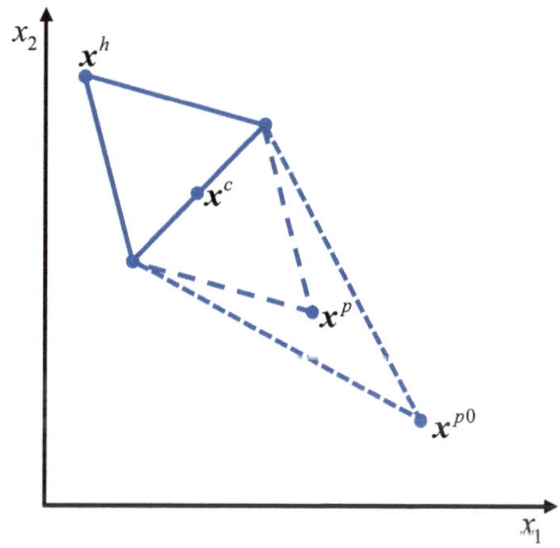

Fig. 3.11 On the expansion

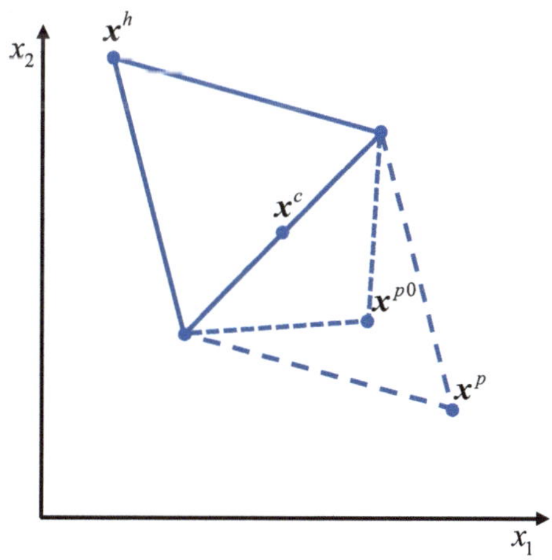

Fig. 3.12 On the outer contraction

- If $f\left(\mathbf{x}_{(i)}^{p0}\right) < f\left(\mathbf{x}_{(i)}^{p}\right)$, then $\mathbf{x}_{(i)}^{h}$ is substituted by $\mathbf{x}_{(i)}^{p0}$.
- If $f\left(\mathbf{x}_{(i)}^{p0}\right) \geq f\left(\mathbf{x}_{(i)}^{p}\right)$, then the so-called **reduction** is performed

$$\begin{aligned}\mathbf{x}_{(i+1)}^{j} &= \mathbf{x}_{(i+1)}^{j} + \gamma\left(\mathbf{x}_{(i+1)}^{l} - \mathbf{x}_{(i+1)}^{j}\right) \\ &= \gamma\mathbf{x}_{(i+1)}^{l} + (1-\gamma)\mathbf{x}_{(i+1)}^{j}, j = 1, 2, \ldots, n+1, j \neq l \\ \gamma &\in (0, 1)\end{aligned} \tag{3.35}$$

The reduction (3.35) is analogous to (3.29), but all the vertices move towards the "best" vertex $\mathbf{x}_{(i)}^{l}$ (instead of $\mathbf{x}_{(i)}^{q}$, see Fig. 3.8) in this method. The standard setting of the reduction coefficient is $\gamma = 0.5$.

The last strategy is performed if the worst possible case happens, i.e., $f\left(\mathbf{x}_{(i)}^{p}\right) \geq f\left(\mathbf{x}_{(i)}^{h}\right)$. The so-called **inner contraction** reads

$$\begin{aligned}\mathbf{x}_{(i)}^{p0} &= \mathbf{x}_{(i)}^{c} + \delta\left(\mathbf{x}_{(i)}^{h} - \mathbf{x}_{(i)}^{c}\right) = \delta\mathbf{x}_{(i)}^{h} + (1-\delta)\mathbf{x}_{(i)}^{c} \\ \delta &\in (0, 1)\end{aligned} \tag{3.36}$$

It geometrically means that the flipped point returns inside the current simplex shape in the direction of vector $\mathbf{x}_{(i)}^{c} - \mathbf{x}_{(i)}^{p}$, giving rise to $\mathbf{x}_{(i)}^{p0}$, see Fig. 3.13.

Again, $f\left(\mathbf{x}_{(i)}^{p0}\right)$ is evaluated:

- If $f\left(\mathbf{x}_{(i)}^{p0}\right) < f\left(\mathbf{x}_{(i)}^{h}\right)$, then $\mathbf{x}_{(i)}^{h}$ is substituted by $\mathbf{x}_{(i)}^{p0}$.
- If $f\left(\mathbf{x}_{(i)}^{p0}\right) \geq f\left(\mathbf{x}_{(i)}^{h}\right)$, reduction (3.35) is made.

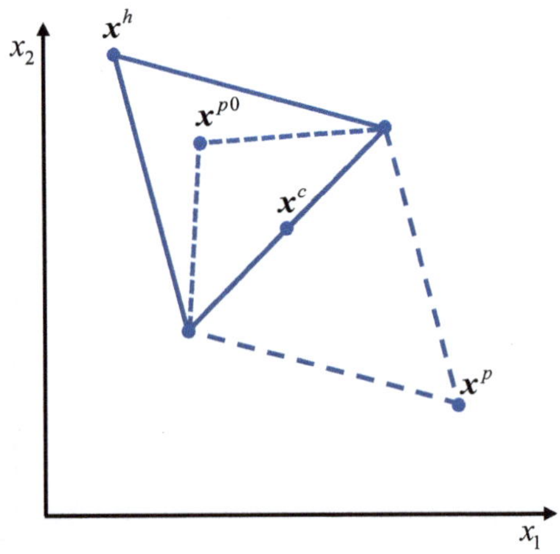

Fig. 3.13 On the inner contraction

A run of the algorithm can geometrically be imagined as a sequence of lengthening and shortening of the simple shape when its moving in the parameter space. The overall simplex size should asymptotically tend to zero. It reminds a move of an amoeba in $\mathbb{R}^2$. The irregularity of the simplex preserves the algorithm from any of the loopings described in Sect. 3.1.2.1.

A termination condition can, e.g., be the average simplex edge length

$$\overline{\Delta} = \frac{1}{n+1}\left(\left\|\mathbf{x}_{(i)}^{n+1} - \mathbf{x}_{(i)}^{1}\right\| + \sum_{j=1}^{n}\left\|\mathbf{x}_{(i)}^{j+1} - \mathbf{x}_{(i)}^{j}\right\|\right) \tag{3.37}$$

Note that $\|\cdot\|$ in (3.37) expresses the Euclidean norm.

The complete algorithm of the Nelder–Mead method follows.

Algorithm 3.5 (The irregular or Nelder–Mead simplex method)

1. Let $f(\mathbf{x})$ be given. Select the initial point $\mathbf{x}_{(1)}^{1}$, the cathetus length Δ for the initial simplex, the terminating simplex edge length Δ_ε and the expansion, outer contraction, reduction, and inner contraction factors $\alpha > 1$, $\beta, \gamma, \delta \in (0, 1)$, respectively. Initialize the iteration steps counter $i = 1$.
2. Construct the initial irregular simple; e.g., using (3.31).
3. Evaluate function values at vertices $f\left(\mathbf{x}_{(i)}^{j}\right)$, $j = 1, 2, \ldots n+1$ and find $\mathbf{x}_{(i)}^{l}$, $\mathbf{x}_{(i)}^{s}$, and $\mathbf{x}_{(i)}^{h}$ as per (3.32).
4. Compute center $\mathbf{x}_{(i)}^{c}$ using (3.26) and perform flipping (3.27) of $\mathbf{x}_{(i)}^{h}$ to get the flipped point $\mathbf{x}_{(i)}^{p}$.
5. If $f\left(\mathbf{x}_{(i)}^{l}\right) \leq f\left(\mathbf{x}_{(i)}^{p}\right) \leq f\left(\mathbf{x}_{(i)}^{s}\right)$, formally set $\mathbf{x}_{(i+1)}^{h} = \mathbf{x}_{(i)}^{p}$ (with respect to the vertices indexing; i.e., $\mathbf{x}_{(i)}^{p}$ represents a new simplex vertex instead of $\mathbf{x}_{(i)}^{h}$), and go to step 9.
6. If $f\left(\mathbf{x}_{(i)}^{p}\right) < f\left(\mathbf{x}_{(i)}^{l}\right)$, perform expansion (3.33) to get $\mathbf{x}_{(i)}^{p0}$. If $f\left(\mathbf{x}_{(i)}^{p0}\right) < f\left(\mathbf{x}_{(i)}^{l}\right)$, formally set $\mathbf{x}_{(i+1)}^{h} = \mathbf{x}_{(i)}^{p0}$; else, $\mathbf{x}_{(i+1)}^{h} = \mathbf{x}_{(i)}^{p}$. Then, go to step 9.
7. If $f\left(\mathbf{x}_{(i)}^{s}\right) < f\left(\mathbf{x}_{(i)}^{p}\right) < f\left(\mathbf{x}_{(i)}^{h}\right)$, perform outer contraction (3.34) yielding $\mathbf{x}_{(i)}^{p0}$. If $f\left(\mathbf{x}_{(i)}^{p0}\right) < f\left(\mathbf{x}_{(i)}^{p}\right)$, formally set $\mathbf{x}_{(i+1)}^{h} = \mathbf{x}_{(i)}^{p0}$; else, reduce the simple as per (3.35). Then, go to step 9.
8. If $f\left(\mathbf{x}_{(i)}^{p}\right) \geq f\left(\mathbf{x}_{(i)}^{h}\right)$, perform inner contraction (3.36) resulting in $\mathbf{x}_{(i)}^{p0}$. If $f\left(\mathbf{x}_{(i)}^{p0}\right) < f\left(\mathbf{x}_{(i)}^{h}\right)$ formally set $\mathbf{x}_{(i+1)}^{h} = \mathbf{x}_{(i)}^{p0}$; else, reduce the simple via (3.35). Then, go to step 9.
9. Compute the average simplex edge length $\overline{\Delta}$ using (3.37). If $\overline{\Delta} \geq \Delta_\varepsilon$, set $i = i+1$ and go to step 3; else, stop and return $\mathbf{x}^*$ via (3.30).

A sketch of the corresponding Matlab code (for a 2-dimensional case again) follows.

```
% INPUTS
syms f x1 x2
f = input('Enter the objective function: ');
x_init = input('Enter the initial point [x1 x2]: ');
Delta = input('Enter the cathetus length for the initial simplex:
');
Delta_e = input('Enter the desired accuracy (average final simplex
edge length): ');
alpha = input('Enter the simplex expansion factor (alpha > 1): ');
beta = input('Enter the simplex outer contraction factor (0 < beta <
1): ');
gamma = input('Enter the simplex reduction factor (0 < gamma < 1):
');
delta = input('Enter the simplex inner contraction factor (0 <
delta < 1): ');

% INIT
n = 2;
i = 1;
X(1,1:n) = x_init;
F(1) = eval(subs(f,[x1 x2],X(1,:)));

I = eye(n);
for j = 2:1:(n+1)
   X(j,1:n) = x_init + Delta*I(j-1,:);
   F(j) = eval(subs(f,[x1 x2],X(j,:)));
end

% ITERATIONS
while Delta >= Delta_e
   % VERTICES ORDERING
   [F_sort, i_sort] = sort(F);
   ih = i_sort(end); xh = X(ih,1:n); fh=F_sort(end);
   fs=F_sort(end-1);
   fl=F_sort(1);

   % FLIPPING
   if ih==1
      xc = sum(X(2:end,1:n))/n;
   elseif ih==(n+1)
      xc = sum(X(1:ih-1,1:n))/n;
   else
      xc = sum(X([1:ih-1,ih+1:end],1:n))/n;
   end
   xp = 2*xc - xh;
   fp = eval(subs(f,[x1 x2],xp));

   % STRATEGIES
   if (fp >= fl) && (fp <= fs) % SIMPLE VERTEX SUBSTITUTION
      X(ih,1:n) = xp;
      F(ih) = eval(subs(f,[x1 x2],xp));
```

```
    elseif fp < fl % EXPANSION
      xp0 = alpha*xp + (1-alpha)*xc;
      fxp0 = eval(subs(f,[x1 x2],xp0));
      if fxp0 < fl
        X(ih,1:n) = xp0;
        F(ih) = eval(subs(f,[x1 x2],xp0));
      else
        X(ih,1:n) = xp;
        F(ih) = eval(subs(f,[x1 x2],xp));
      end
    elseif (fp > fs) && (fp < fh) % OUTER CONTRACTION
      xp0 = beta*xp + (1-beta)*xc;
      fxp0 = eval(subs(f,[x1 x2],xp0));
      if fxp0 < fp
        X(ih,1:n) = xp0;
        F(ih) = eval(subs(f,[x1 x2],xp0));
      else   % REDUCTION
        il = i_sort(1);
        for j = 1:1:(n+1)
          X(j,1:n) = gamma*X(il,1:n) + (1-gamma)*X(j,1:n);
          F(j) = eval(subs(f,[x1 x2],X(j,:)));
        end
      end
    else                        % INNER CONTRACTION
      xp0 = delta*xp + (1-delta)*xc;
      fxp0 = eval(subs(f,[x1 x2],xp0));
      if fxp0 < fh
        X(ih,1:n) = xp0;
        F(ih) = eval(subs(f,[x1 x2],xp0));
      else % REDUCTION
        il = i_sort(1);
        for j = 1:1:(n+1)
          X(j,1:n) = gamma*X(il,1:n) + (1-gamma)*X(j,1:n);
          F(j) = eval(subs(f,[x1 x2],X(j,:)));
        end
      end
    end

    % AVERAGE SIMPLEX EDGE LENGTH
    Delta=0;
    for j = 1:1:n
      Delta = Delta + norm(X(j+1,1:n)-X(j,1:n));
    end
    Delta = (Delta + norm(X(1,1:n)-X(n+1,1:n)))/(n+1);
end

% OUTPUT
X(i_sort(1),1:n)
```

Example 3.5 *Task*: Find a minimum of $f(\mathbf{x}) = f(x_1, x_2) = 3x_1^2 - 2x_1x_2 + x_2^2 + 4x_1 + 3x_2 + 1$ on $\mathbb{R}^2$ using the irregular simplex (Nelder–Mead) method. Let the initial point be $\mathbf{x}_{(1)}^1 = [0, 0]^T$, the initial simplex edge length $\Delta = 1$, and the expansion, outer contraction, reduction, and inner contraction factors $\alpha = 2, \beta = \gamma = \delta = 0.5$, respectively. Choose the fixed number of $N = 15$ iteration steps. What is the eventual average simplex edge length?

Solution: Two iteration steps of the algorithm are demonstrated in detail, the rest of the calculation will be presented in a table.

Set $i = 1$ and generate the initial simplex using (3.31), starting from $\mathbf{x}^1_{(1)} = [0, 0]^T$

$$\begin{aligned}\mathbf{x}^1_{(1)} &= [0, 0]^T \\ \mathbf{x}^2_{(1)} &= \mathbf{x}^1_{(1)} + \Delta\mathbf{e}_1 = [0, 0]^T + [1, 0]^T = [1, 0]^T \\ \mathbf{x}^3_{(1)} &= \mathbf{x}^1_{(1)} + \Delta\mathbf{e}_2 = [0, 0]^T + [0, 1]^T = [0, 1]^T\end{aligned}$$

The points with the smallest, second highest, and highest functional values are searched

$$f\left(\mathbf{x}^1_{(1)} = \mathbf{x}^l_{(1)}\right) = 1, f\left(\mathbf{x}^2_{(1)} = \mathbf{x}^h_{(1)}\right) = 8, f\left(\mathbf{x}^3_{(1)} = \mathbf{x}^s_{(1)}\right) = 5$$

Reflect the vertex $\mathbf{x}^h_{(1)}$ through the center point $\mathbf{x}^c_{(1)}$ (similarly to the flipping operation in the regular simplex method)

$$\mathbf{x}^c_{(1)} = \frac{\mathbf{x}^l_{(1)} + \mathbf{x}^s_{(1)}}{2} = [0, 0.5]^T, \mathbf{x}^p_{(1)} = 2\mathbf{x}^c_{(1)} - \mathbf{x}^h_{(1)} = [-1, 1]^T$$

Compute that $f\left(\mathbf{x}^p_{(1)}\right) = 6$, i.e., $f\left(\mathbf{x}^s_{(1)}\right) < f\left(\mathbf{x}^p_{(1)}\right) < f\left(\mathbf{x}^h_{(1)}\right)$. Therefore, the outer contraction is performed

$$\mathbf{x}^{p0}_{(1)} = \mathbf{x}^c_{(1)} + \beta\left(\mathbf{x}^p_{(1)} - \mathbf{x}^c_{(1)}\right) = [-0.5, 0.75]^T, \beta = 0.5$$

As $f\left(\mathbf{x}^{p0}_{(1)}\right) = 3.3125 < 6 = f\left(\mathbf{x}^p_{(1)}\right)$, the contraction has been successful, and hence, the new simplex vertex reads $\mathbf{x}^2_{(2)} = \mathbf{x}^{p0}_{(1)}$.

The iteration counter is incremented to $i = 2$, and it is evaluated that $f\left(\mathbf{x}^1_{(2)}\right) = 1, f\left(\mathbf{x}^2_{(2)}\right) = 3.321, f\left(\mathbf{x}^3_{(2)}\right) = 5$, i.e., $\mathbf{x}^l_{(2)} = \mathbf{x}^1_{(2)} = [0, 0]^T$, $\mathbf{x}^s_{(2)} = \mathbf{x}^2_{(2)} = [-0.5, 0.75]^T$, $\mathbf{x}^h_{(2)} = \mathbf{x}^3_{(2)} = [0, 1]^T$. The vertex $\mathbf{x}^h_{(2)}$ is reflected, yielding

$$\mathbf{x}^c_{(2)} = \frac{\mathbf{x}^l_{(2)} + \mathbf{x}^s_{(2)}}{2} = [-0.25, 0.375]^T, \mathbf{x}^p_{(2)} = 2\mathbf{x}^c_{(2)} - \mathbf{x}^h_{(2)} = [-0.5, -0.25]^T$$

Since $f\left(\mathbf{x}^p_{(2)}\right) = -1.1875 < 1 = f\left(\mathbf{x}^l_{(2)}\right)$, the expansion

$$\mathbf{x}^{p0}_{(2)} = \mathbf{x}^c_{(2)} + \alpha\left(\mathbf{x}^p_{(2)} - \mathbf{x}^c_{(2)}\right) = [-0.75, -0.875]^T, \alpha = 2$$

is performed. The function value at the expanded point $f\left(\mathbf{x}_{(2)}^{p0}\right) = -3.4844$ is better than $f\left(\mathbf{x}_{(2)}^{l}\right) = 1$ (i.e., $f\left(\mathbf{x}_{(2)}^{p0}\right) < f\left(\mathbf{x}_{(2)}^{l}\right)$); hence, $\mathbf{x}_{(2)}^{h}$ is substituted by $\mathbf{x}_{(2)}^{p0}$, which is formally expressed as $\mathbf{x}_{(3)}^{3} = \mathbf{x}_{(2)}^{p0}$.

The remaining procedure is summarized in Table 3.5. The "operation" column indicates how the flipped point is modified: no modification (0), expansion (E), outer contraction (OC), inner contraction (IC), or reduction (R).

As can be seen from Table 3.5, after the expansion phase, a sequence of inner contractions reduces the simplex size in the neighborhood of the analytic extreme point $\mathbf{x}^* = [-1.75, -3.25]^T$. After the 15th iteration, the eventual simplex has vertices at $\mathbf{x}^1 = [-1.7385, -3.2843]^T$, $\mathbf{x}^2 = [-1.7522, -3.2358]^T$, $\mathbf{x}^3 = [-1.774, -3.3013]^T$.

Either the vertex with a minimum function value (i.e., $\tilde{\mathbf{x}}^* = \mathbf{x}^2 = \mathbf{x}^l, f\left(\mathbf{x}^2\right) = -7.3747$) or the simplex center $\tilde{\mathbf{x}}^* = [-1.7549, -3.2738]^T$ (with $f(\tilde{\mathbf{x}}^*) = -7.3746$) can be taken as the minimum estimate. Simplex edges are of lengths 0.0504, 0.0394, 0.069. Their average value is 0.0529, which is significantly lower than for a regular simplex in Example 3.4. Hence, the Nelder–Mead method works faster and it is more accurate in this example, yielding the absolute minimum estimation error of $[0.0022, 0.0142]^T$ or $[0.0049, 0.0238]^T$, respectively.

The third method of this section inspects the variable space via orthogonal jumps with a single-dimensional suboptimization.

3.1.2.3 Cyclic Parameter Swapping (Gauss–Seidel) Method

The presented method is named after Carl Friedrich Gauss (1777–1855) and Philipp Ludwig von Seidel (1821–1896). It represents, in principle, a simple and intuitive procedure combining a successive one-dimensional projection of $f(\mathbf{x})$ on $p(\lambda)$, $\lambda \in \mathbb{R}$ and the analytic extremum λ^* seeking for $p(\lambda)$. From this viewpoint, it stands between comparative and gradient-based methods.

The basic idea is to fix all the variables (parameters) at their current values $x_j, j = 1, 2, \ldots, n, j \neq k$, except for a single parameter that represents a variable $x_k \rightarrow \lambda_k$ for a particular $k \in \{1, 2, \ldots, n\}$. Every iteration step has n substeps. In each substep, the optimized variable λ_k^* is changed (swapped) so that k is incremented (i.e., $k = 1, 2, \ldots, n$).

Geometrically, the algorithm can be thought of as a *rectangular* movement along the x_k axis in the space of variables; see Fig. 3.14. A global extremum in the x_k-direction is found analytically. In other words, rectangular one-dimensional cuts are made.

In some cases, it is possible to express the relation

$$\lambda_k^* = \left[x_1, x_2, \ldots, x_{k-1}, x_{k+1}, \ldots, x_n\right]^T \tag{3.38}$$

Table 3.5 Irregular simplex method—Example 3.5

i	$\mathbf{x}^1_{(i)}$ / $f\left(\mathbf{x}^1_{(i)}\right)$	$\mathbf{x}^2_{(i)}$ / $f\left(\mathbf{x}^2_{(i)}\right)$	$\mathbf{x}^3_{(i)}$ / $f\left(\mathbf{x}^3_{(i)}\right)$	$\mathbf{x}^p_{(i)}$ / $f\left(\mathbf{x}^p_{(i)}\right)$	$\mathbf{x}^{p0}_{(i)}$ / $f\left(\mathbf{x}^{p0}_{(i)}\right)$	Operation
1	$\begin{bmatrix} 0 \\ 0 \end{bmatrix}$	$\begin{bmatrix} 1 \\ 0 \end{bmatrix}$	$\begin{bmatrix} 0 \\ 1 \end{bmatrix}$	$\begin{bmatrix} -1 \\ 1 \end{bmatrix}$	$\begin{bmatrix} -0.5 \\ 0.75 \end{bmatrix}$	OC
	1	8	5	6	3.3125	
2	$\begin{bmatrix} 0 \\ 0 \end{bmatrix}$	$\begin{bmatrix} -0.5 \\ 0.75 \end{bmatrix}$	$\begin{bmatrix} 0 \\ 1 \end{bmatrix}$	$\begin{bmatrix} -0.5 \\ -0.25 \end{bmatrix}$	$\begin{bmatrix} -0.75 \\ -0.875 \end{bmatrix}$	E
	1	3.3125	5	− 1.1875	− 3.4844	
3	$\begin{bmatrix} 0 \\ 0 \end{bmatrix}$	$\begin{bmatrix} -0.5 \\ -0.25 \end{bmatrix}$	$\begin{bmatrix} -0.75 \\ -0.875 \end{bmatrix}$	$\begin{bmatrix} -0.25 \\ -1.625 \end{bmatrix}$	N/A	0
	1	3.3125	− 3.4844	− 2.8594	N/A	
4	$\begin{bmatrix} 0 \\ 0 \end{bmatrix}$	$\begin{bmatrix} -0.25 \\ -1.625 \end{bmatrix}$	$\begin{bmatrix} -0.75 \\ -0.875 \end{bmatrix}$	$\begin{bmatrix} -1 \\ -2.5 \end{bmatrix}$	$\begin{bmatrix} -1.5 \\ -3.75 \end{bmatrix}$	E
	1	− 2.8594	− 3.4844	− 6.25	− 6.6875	
5	$\begin{bmatrix} -1.5 \\ -3.75 \end{bmatrix}$	$\begin{bmatrix} -0.25 \\ -1.625 \end{bmatrix}$	$\begin{bmatrix} -0.75 \\ -0.875 \end{bmatrix}$	$\begin{bmatrix} -2 \\ -3 \end{bmatrix}$	$\begin{bmatrix} -2.875 \\ -3.6875 \end{bmatrix}$	E
	− 6.6875	− 2.8594	− 3.4844	− 7	− 4.3711	
6	$\begin{bmatrix} -1.5 \\ -3.75 \end{bmatrix}$	$\begin{bmatrix} -2 \\ -3 \end{bmatrix}$	$\begin{bmatrix} -0.75 \\ -0.875 \end{bmatrix}$	$\begin{bmatrix} -2.75 \\ -5.875 \end{bmatrix}$	$\begin{bmatrix} -1.25 \\ -2.125 \end{bmatrix}$	IC
	− 6.6875	− 7	− 3.4844	− 2.7344	− 6.4844	
7	$\begin{bmatrix} -1.5 \\ -3.75 \end{bmatrix}$	$\begin{bmatrix} -2 \\ -3 \end{bmatrix}$	$\begin{bmatrix} -1.25 \\ -2.125 \end{bmatrix}$	$\begin{bmatrix} -2.25 \\ -4.625 \end{bmatrix}$	$\begin{bmatrix} -1.5 \\ -2.75 \end{bmatrix}$	IC
	− 6.6875	− 7	− 6.4844	− 6.1094	− 7.1875	
8	$\begin{bmatrix} -1.5 \\ -3.75 \end{bmatrix}$	$\begin{bmatrix} -2 \\ -3 \end{bmatrix}$	$\begin{bmatrix} -1.5 \\ -2.75 \end{bmatrix}$	$\begin{bmatrix} -2 \\ -2 \end{bmatrix}$	$\begin{bmatrix} -1.625 \\ -3.3125 \end{bmatrix}$	IC
	− 6.6875	− 7	− 7.1875	− 5	− 7.3086	
9	$\begin{bmatrix} -1.625 \\ -3.3125 \end{bmatrix}$	$\begin{bmatrix} -2 \\ -3 \end{bmatrix}$	$\begin{bmatrix} -1.5 \\ -2.75 \end{bmatrix}$	$\begin{bmatrix} -1.125 \\ -3.0625 \end{bmatrix}$	$\begin{bmatrix} -1.7813 \\ -3.0156 \end{bmatrix}$	IC
	− 7.3086	− 7	− 7.1875	− 6.4023	− 7.3025	
10	$\begin{bmatrix} -1.625 \\ -3.3125 \end{bmatrix}$	$\begin{bmatrix} -1.7813 \\ -3.0156 \end{bmatrix}$	$\begin{bmatrix} -1.5 \\ -2.75 \end{bmatrix}$	$\begin{bmatrix} -1.9063 \\ -3.5781 \end{bmatrix}$	$\begin{bmatrix} -1.8047 \\ -3.3711 \end{bmatrix}$	OC
	− 7.3086	− 7.3025	− 7.1875	− 7.2966	− 7.3646	

(continued)

Table 3.5 (continued)

i	$\mathbf{x}^1_{(i)}$ / $f\left(\mathbf{x}^1_{(i)}\right)$	$\mathbf{x}^2_{(i)}$ / $f\left(\mathbf{x}^2_{(i)}\right)$	$\mathbf{x}^3_{(i)}$ / $f\left(\mathbf{x}^3_{(i)}\right)$	$\mathbf{x}^p_{(i)}$ / $f\left(\mathbf{x}^p_{(i)}\right)$	$\mathbf{x}^{p0}_{(i)}$ / $f\left(\mathbf{x}^{p0}_{(i)}\right)$	Operation
11	$\begin{bmatrix}-1.625\\-3.3125\end{bmatrix}$	$\begin{bmatrix}-1.7813\\-3.0156\end{bmatrix}$	$\begin{bmatrix}-1.8047\\-3.3711\end{bmatrix}$	$\begin{bmatrix}-1.6484\\-3.668\end{bmatrix}$	$\begin{bmatrix}-1.748\\-3.1787\end{bmatrix}$	IC
	−7.3086	−7.3025	−7.3646	−7.0845	−7.3702	
12	$\begin{bmatrix}-1.625\\-3.3125\end{bmatrix}$	$\begin{bmatrix}-1.7813\\-3.0156\end{bmatrix}$	$\begin{bmatrix}-1.8047\\-3.3711\end{bmatrix}$	$\begin{bmatrix}-1.9277\\-3.2373\end{bmatrix}$	$\begin{bmatrix}-1.7007\\-3.2937\end{bmatrix}$	IC
	−7.3086	−7.3702	−7.3646	−7.2756	−7.3615	
13	$\begin{bmatrix}-1.7007\\-3.2937\end{bmatrix}$	$\begin{bmatrix}-1.748\\-3.1787\end{bmatrix}$	$\begin{bmatrix}-1.8047\\-3.3711\end{bmatrix}$	$\begin{bmatrix}-1.8521\\-3.2561\end{bmatrix}$	$\begin{bmatrix}-1.7385\\-3.2843\end{bmatrix}$	IC
	−7.3615	−7.3702	−7.3646	−7.345	−7.3726	
14	$\begin{bmatrix}-1.7385\\-3.2843\end{bmatrix}$	$\begin{bmatrix}-1.748\\-3.1787\end{bmatrix}$	$\begin{bmatrix}-1.8047\\-3.3711\end{bmatrix}$	$\begin{bmatrix}-1.6819\\-3.0919\end{bmatrix}$	$\begin{bmatrix}-1.774\\-3.3013\end{bmatrix}$	IC
	−7.3726	−7.3702	−7.3646	−7.3576	−7.3731	
15	$\begin{bmatrix}-1.7385\\-3.2843\end{bmatrix}$	$\begin{bmatrix}-1.748\\-3.1787\end{bmatrix}$	$\begin{bmatrix}-1.774\\-3.3013\end{bmatrix}$	$\begin{bmatrix}-1.7645\\-3.4069\end{bmatrix}$	$\begin{bmatrix}-1.7522\\-3.2358\end{bmatrix}$	IC
	−7.3726	−7.3702	−7.3731	−7.3543	−7.3747	

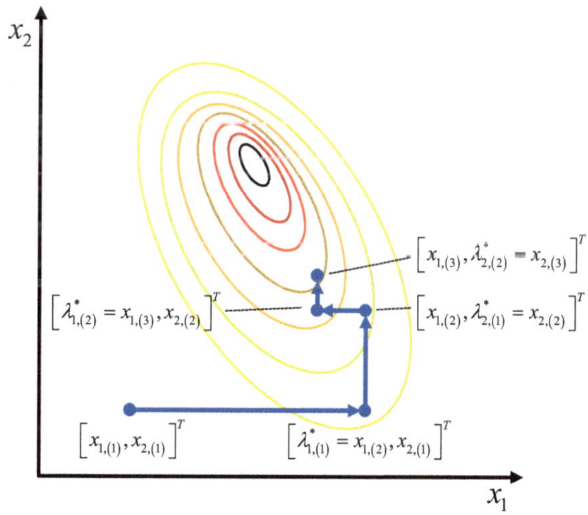

Fig. 3.14 The principle of the cyclic parameter swapping method with indicated contours of $f(\mathbf{x})$

analytically. It means that it is not necessary to solve the one-dimensional optimization problem in every single iteration step, but one can evaluate λ_k^* directly based on the fixed right-hand side of (3.38).

It should hold $\lim_{i\to 0} \lambda_{k,(i)}^* = 0$, for $k = 1, 2, \ldots, n$. Hence, a run of the iterations can be terminated if

$$\left\| \begin{bmatrix} \lambda_{1,(i)}^* \\ \lambda_{2,(i)}^* \\ \vdots \\ \lambda_{n,(i)}^* \end{bmatrix} - \begin{bmatrix} \lambda_{1,(i-1)}^* \\ \lambda_{2,(i-1)}^* \\ \vdots \\ \lambda_{n,(i-1)}^* \end{bmatrix} \right\| = \left\| \mathbf{x}_{(i)} - \mathbf{x}_{(i-1)} \right\| < \varepsilon \tag{3.39}$$

for some $i > 1$ and a selected accuracy $\varepsilon > 0$.

The method can quickly get stuck at a local extremum in the case of a multimodal objective function. Compared to the simplex methods (see Sects. 3.1.2.1 and 3.1.2.2), it has a worse ability to inspect the variable space.

The complete algorithm and a corresponding Matlab code (for the minimization problem) for a two-dimensional case follow again.

Algorithm 3.6 (The cyclic parameter swapping or Gauss–Seidel method)

1. Let a differentiable $f(\mathbf{x})$ be given. Select the initial point $\mathbf{x}_{(1)}^1$ and the desired accuracy $\varepsilon > 0$. Initialize the iteration steps counter $i = 1$ and the counter of substeps (i.e., the variable pass counter) $k = 1$.
2. Construct one-dimensional function

$$p(\lambda_{k,(i)}) := f(\mathbf{x})\big|_{\mathbf{x}=[x_{1,(i)},\ldots,x_{k-1,(i)},\lambda_{k,(i)},x_{k+1,(i)}\ldots,x_{n,(i)}]^T} \tag{3.40}$$

3. Solve the optimization problem

$$\lambda_{k,(i)}^* = \arg\operatorname{extr} p(\lambda_{k,(i)}) \tag{3.41}$$

   using the technique described in Sect. 2.1.
4. Set $x_{k,(i)} = \lambda_{k,(i)}^*$. If $k < n$, increment $k = k + 1$ and go to step 2; else, set $k = 1$, and go to step 5.
5. Set $i = i + 1$ and $\mathbf{x}_{(i)} = \left[\lambda_{1,(i-1)}^*, \lambda_{2,(i-1)}^*, \ldots, \lambda_{n,(i-1)}^*\right]^T$. Perform termination test (3.39). If (3.39) does not hold, go to step 2; else, stop and return $\mathbf{x}^* = \mathbf{x}_{(i)}$.

```
% INPUTS
syms f x1 x2
f = input('Enter the objective function: ');
x_init = input('Enter the initial point [x1 x2]: ');
e = input('Enter the desired accuracy: ');


% INIT
```

```
n = 2;
i = 1;
k = 1;
x_prev = x_init;
x = x_init;
dx = inf;

% ITERATIONS
while dx >= e
   % SUBSTEPS
   while k <= n
      if k == 1
         p = subs(f,x2,x(2));
         % ONE-DIMENSIONAL OPTIMIZATION
         dp = diff(p,x1);
         X0 = solve(dp==0);
         n_X0 = length(X0);
         f_min=inf;
         for j = 1:1:n_X0
            l = 2;
            while eval(subs(diff(dp,x1),x1,X0(j))==0)
               dp = diff(dp,x1);
               l = l+1;
            end
            if (rem(l,2)==0) && eval(subs(diff(dp,x1),x1,X0(j))>0)
               if eval(subs(p,x1,X0(j))) < f_min
                  x(1) = X0(j); % A BETTER SOLUTION FOUND
               end
            end
         end
      else
         p = subs(f,x1,x(1));
         % ONE-DIMENSIONAL OPTIMIZATION
         dp = diff(p,x2);
         X0 = solve(dp==0);
         n_X0 = length(X0);
         f_min=inf;
         for j = 1:1:n_X0
            l = 2;
            while eval(subs(diff(dp,x2),x2,X0(j))==0)
               dp = diff(dp,x2);
               l = l+1;
            end
            if (rem(l,2)==0) && eval(subs(diff(dp,x2),x2,X0(j))>0)
               if eval(subs(p,x2,X0(j))) < f_min
                  x(2) = X0(j); % A BETTER SOLUTION FOUND
               end
            end
         end
      end
      X0 = 0;
      k = k+1;
   end
   dx = norm(x-x_prev);
```

```
    x_prev = x;
    k = 1;
end

% OUTPUT
x
```

Example 3.6 *Task*: Find a minimum of $f(\mathbf{x}) = f(x_1, x_2) = 3x_1^2 - 2x_1x_2 + x_2^2 + 4x_1 + 3x_2 + 1$ on $\mathbb{R}^2$ using the cyclic parameter swapping (Gauss–Seidel) method. Let the initial estimate be $\mathbf{x}_{(1)} = [0, 0]^T$ and the desired accuracy $\varepsilon = 0.01$.

Solution: Analogously to the preceding examples, two iteration steps are presented in detail, and the rest of the computations are displayed in a table.

Define $p(\lambda_1)$ generally as

$$p(\lambda_1) = f(\mathbf{x})|_{\mathbf{x}=[\lambda_1, x_2]^T} = 3\lambda_1^2 - 2\lambda_1 x_2 + x_2^2 + 4\lambda_1 + 3x_2 + 1$$

Recall that x_2 has a fixed value, while λ_1 represents a variable. As $p(\lambda_1)$ is differentiable (i.e., a 2nd order polynomial), its minimum can easily be calculated, instead of computing $\lambda^*_{1,(i)}$ at each iteration step for particular x_2. Hence, following Theorem 2. 1, it can be calculated:

$$\frac{dp(\lambda_1)}{d\lambda_1} = 6\lambda_1 - 2x_2 + 4 \overset{!}{=} 0$$
$$\Rightarrow \lambda_{1,0} = \frac{x_2 - 2}{3}$$

As $d^2p(\lambda_1)/d\lambda_1^2 = p''(\lambda_1) = 6 > 0$, the found stationary point $\lambda_{1,0}$ represents a (global) minimum.

Analogously for $p(\lambda_2)$, one has

$$p(\lambda_2) = f(\mathbf{x})|_{\mathbf{x}=[x_1, \lambda_2]^T} = 3x_1^2 - 2x_1\lambda_2 + \lambda_2^2 + 4x_1 + 3\lambda_2 + 1$$
$$\frac{dp(\lambda_2)}{d\lambda_2} = -2x_1 + 2\lambda_2 + 3 \overset{!}{=} 0$$
$$\Rightarrow \lambda_{2,0} = \frac{2x_1 - 3}{2}$$
$$\frac{d^2p(\lambda_2)}{d\lambda_2^2} = 2 > 0$$

Now, starting from $\mathbf{x}_{(1)} = [0, 0]^T$, set $k = 1$ and compute

$$\lambda^*_{1,(1)} = \frac{x_{2,(1)} - 2}{3} = \frac{0-2}{3} = -\frac{2}{3} \Rightarrow \left[\lambda^*_{1,(1)}, x_{2,(1)}\right]^T = \left[-\frac{2}{3}, 0\right]^T$$

and at this updated point, set $k = 2$, and find $\lambda^*_{2,(1)}$ via

$$\lambda^*_{2,(1)} = \frac{2\lambda^*_{1,(1)} - 3}{2} = \frac{2(-\frac{2}{3}) - 3}{2} = -\frac{13}{6}$$
$$\Rightarrow \mathbf{x}_{(2)} = \left[\lambda^*_{1,(1)}, \lambda^*_{2,(1)}\right]^T = \left[-\frac{2}{3}, -\frac{13}{6}\right]^T$$

The second iteration step ($i = 2$) applies the following computations

$$\lambda^*_{1,(2)} = \frac{x_{2,(2)} - 2}{3} = \frac{-\frac{13}{6} - 2}{3} = -\frac{25}{18} = -1.3889$$
$$\lambda^*_{2,(2)} = \frac{2\lambda^*_{1,(2)} - 3}{2} = \frac{2(-\frac{25}{18}) - 3}{2} = -\frac{26}{9} = -2.8889$$
$$\Rightarrow \mathbf{x}_{(3)} = [-1.3889, -2.8889]^T$$

The corresponding objective function values successively decrease

$$f\left(\mathbf{x}_{(1)}\right) = 1$$
$$f\left(\left[\lambda^*_{1,(1)}, x_{2,(1)}\right]\right) = \frac{-1}{3}$$
$$f\left(\mathbf{x}_{(2)}\right) = -5.0278$$
$$f\left(\left[\lambda^*_{1,(2)}, x_{2,(2)}\right]\right) = -6.5926$$
$$f\left(\mathbf{x}_{(3)}\right) = -7.1142$$

Note that the Euclidean norms of the solution changes read

$$\left\|\mathbf{x}_{(2)} - \mathbf{x}_{(1)}\right\| = \left\|\begin{bmatrix} -\frac{2}{3} \\ -\frac{13}{6} \end{bmatrix} - \begin{bmatrix} 0 \\ 0 \end{bmatrix}\right\| = 2.2669$$
$$\left\|\mathbf{x}_{(3)} - \mathbf{x}_{(2)}\right\| = \left\|\begin{bmatrix} -\frac{25}{18} \\ -\frac{26}{9} \end{bmatrix} - \begin{bmatrix} -\frac{2}{3} \\ -\frac{13}{6} \end{bmatrix}\right\| = 1.0214$$

The rest of the example is summarized in Table 3.6.

The eventual global minimum estimate is at $\mathbf{x}^* = [-1.7485, -3.2485]^T$ after the 8th iteration. It is clear that this method converges faster than simplex methods for the given example. This is due to the fact that the objective function is convex on $\mathbb{R}^2$, since its Hessian reads

$$\mathbf{H}(f(\mathbf{x}^*)) = \begin{pmatrix} 6 & -2 \\ -2 & 6 \end{pmatrix}$$

The corresponding quadratic form with the Hessian is positive definite; see Propositions 2.4 and 2.5.

Table 3.6 Cyclic parameter swapping method—Example 3.6

i	$\mathbf{x}_{(i)}$	$\lambda^*_{1,(i)}$	$\lambda^*_{2,(i)}$	$f(\mathbf{x}_{(i)})$	$\lVert \mathbf{x}_{(i)} - \mathbf{x}_{(i-1)} \rVert$
1	$\begin{bmatrix} 0 \\ 0 \end{bmatrix}$	$-2/3$	$-13/6$	1	–
2	$\begin{bmatrix} -2/3 \\ -13/6 \end{bmatrix}$	$-25/18$	$-26/9$	-5.0278	2.2669
3	$\begin{bmatrix} -25/18 \\ -26/9 \end{bmatrix}$	-1.6296	-3.1296	-7.1142	1.0214
4	$\begin{bmatrix} -1.6296 \\ -3.1296 \end{bmatrix}$	-1.7099	-3.2099	-7.3460	0.3405
5	$\begin{bmatrix} -1.7099 \\ -3.2099 \end{bmatrix}$	-1.7366	-3.2366	-7.3718	0.1135
6	$\begin{bmatrix} -1.7366 \\ -3.2366 \end{bmatrix}$	-1.7455	-3.2455	-7.3746	0.0378
7	$\begin{bmatrix} -1.7455 \\ -3.2455 \end{bmatrix}$	-1.7485	-3.2485	-7.3750	0.0126
8	$\begin{bmatrix} -1.7485 \\ -3.2485 \end{bmatrix}$	-1.7485	-3.2485	-7.3750	0.0042

However, the method may not converge for non-convex (non-concave) functions or stray into a local minimum (maximum). The simplex methods are more robust in those cases. Another drawback is that demanding analytic computation is needed.

3.2 Gradient-Based Methods

It is a well-known fact that the gradient of a function (if it exists) indicates the *direction of steepest growth* of its function values; see Fig. 2.8. This feature is used by the so-called gradient-based iteration methods that are based on the calculation or estimation of the gradient (or derivative in the one-dimensional case), at the point of the current extremum estimate. Various modifications of this principle enable, e.g., less demanding calculation, better convergence, etc.

In general, the principle of gradient iteration methods can be expressed by the relation

$$\begin{aligned} &x_{(i+1)} = x_{(i)} + \lambda_{(i)} f'\left(x_{(i)}\right),\ x \in \mathbb{R} \\ &\mathbf{x}_{(i+1)} = \mathbf{x}_{(i)} + \lambda_{(i)} \operatorname{grad} f\left(\mathbf{x}_{(i)}\right),\ \mathbf{x} \in \mathbb{R}^n \\ &\lambda_{(i)} \in \mathbb{R} \end{aligned} \tag{3.42}$$

for one-dimensional and multiple-dimensional cases, respectively.

Proposition 3.1 *Let* grad $f(\mathbf{x})$ *at some fixed* $\mathbf{x} = \mathbf{x}_{(i)}$ *exist. Then, for a sufficiently small* $\lambda_{(i)} > 0$ *in (3.42), it holds that* $f\left(\mathbf{x}_{(i+1)}\right) > f\left(\mathbf{x}_{(i)}\right)$*. Contrariwise,* $f\left(\mathbf{x}_{(i+1)}\right) < f\left(\mathbf{x}_{(i)}\right)$ *for* $\lambda_{(i)} < 0$*.*

Proposition 3.1 implies that the former option is used to search of a maximum, while the latter is for a minimum. Naturally, it also holds that for a one-dimensional case.

Let two simple one-dimensional methods be presented first. Then, readers are acquainted with several multiple-dimensional methods intended primarily to unconstrained problems.

3.2.1 One-Dimensional Problem

The optimization problem in question has been defined in (2.8). In contrast to gradient-free methods introduced in Sect. 3.1.1, one or more initial points (estimates) are considered instead of a given interval of possible solution candidates.

3.2.1.1 Newton Method

It represents a method suitable for finding the extrema of a *unimodal* and even more suitable *convex* (*concave*) objective function. The iteration relation of the Newton method is based on the function expansion using the Taylor series around the current iteration (extremum estimate).

Let $f(x)$ be differentiable up to the order of two, and the following Taylor series exists at fixed point $x_{(i)}$ in the neighborhood of the function extreme x^*

$$f(x) = f\left(x_{(i)}\right) + f'\left(x_{(i)}\right)\left(x - x_{(i)}\right) + \frac{1}{2}f''\left(x_{(i)}\right)\left(x - x_{(i)}\right)^2 + R\left(\left|x - x_{(i)}\right|\right) \quad (3.43)$$

where $R\left(\left|x - x_{(i)}\right|\right)$ expresses a remainder that is neglected.

Then, if the necessary extremum condition (Proposition 2.1) $f'(x)\big|_{x=\tilde{x}^*} = f'(\tilde{x}^*) \overset{!}{=} 0$ is applied to (3.43), one obtains the extremum estimate

$$\tilde{x}^* = x_{(i)} - \frac{f'\left(x_{(i)}\right)}{f''\left(x_{(i)}\right)} \quad (3.44)$$

Finally, if it is assumed that the required extremum is not reached at one single step but iteratively, (3.44) yields

$$x_{(i+1)} = x_{(i)} - \frac{f'\left(x_{(i)}\right)}{f''\left(x_{(i)}\right)} \tag{3.45}$$

The Newton method works well in the neighborhood of an extremum; however, it is unsuitable for searching global extrema in the case of multimodal or non-convex (non-concave) functions. Namely, it always reaches the nearest (local) extremum based on objective function properties at the initial point $x_{(1)}$. If $f(x)$ is *convex* at $x_{(1)}$, the nearest *minimum* is found, if $f(x)$ is *concave* at $x_{(1)}$, the nearest *maximum* is found.

The rationale for that is as follows. Consider a convex objective function at the initial point. It means that $f''\left(x_{(i)}\right) > 0$ (compare with Propositions 2.2 and 2.5). Hence, iteration rule (3.45) can be expressed as

$$x_{(i+1)} = x_{(i)} + \lambda_{(i)} f'\left(x_{(i)}\right), \ \lambda_{(i)} = -\frac{1}{f''\left(x_{(i)}\right)} < 0 \tag{3.46}$$

In the light of Proposition 3.1, (3.46) yields $x_{(i+1)} < x_{(i)}$.

Another problem can appear if $f(x)$ is "too flat" in the neighborhood of an extremum (i.e., a non-strict local extremum exists, see Fig. 2.3). Then, $f'(x) \approx 0$ and $f''(x) \approx 0$ at some x near by x^* (see also Example 2.2), and the computation of (3.45) can suffer from numerical errors.

A termination condition can, e.g., be simply given by

$$\left|x_{(i+1)} - x_{(i)}\right| < \varepsilon \tag{3.47}$$

The complete algorithm follows.

Algorithm 3.7 (The one-dimensional Newton method)

1. Let objective function $f(x)$ (differentiable up to the order of two) and the initial extremum estimate $x_{(1)}$ be given. Select $\varepsilon > 0$. Set $i = 1$.
2. Calculate $f'(x)$ and $f''(x)$.
3. Compute $x_{(i+1)}$ via (3.45).
4. If (3.47) does not hold, set $i = i + 1$ and go to step 3. Otherwise, stop and return $x^* = x_{(i+1)}$.

It is worth noting that a simple strategy for approaching the desired nearest local extremum can be applied in Step 3. The modification of Algorithm 3.7 is based on the altering of $\lambda_{(i)} = -1/f''\left(x_{(i)}\right)$ whenever its value (sign) is insufficient. Namely, if a *minimum*-search task is given and $\lambda_{(i)} > 0$, then one can make the following update

$$\lambda_{(i)} = \lambda_{(i)} - \delta, \ \delta > 0 \tag{3.48}$$

The value of δ is selected so that $\lambda_{(i)} < 0$. In the case of a *maximum*-search task, (3.48) becomes $\lambda_{(i)} = \lambda_{(i)} + \delta, \ \delta > 0$, to get $\lambda_{(i)} > 0$.

A possible Matlab code of Algorithm 3.7 and a corresponding example follow.

```
% INPUTS
syms f x
f = input('Enter the objective function: ');
x_init = input('Enter the initial point (extremum guess): ');
e = input('Enter the desired accuracy: ');

% INIT
df = diff(f,x);
d2f = diff(df,x);
xi = x_init;
xii = x_init + 2*e;
dx = abs(xii-xi);

% ITERATIONS
while dx >= e
   xii = xi - eval(subs(df,x,xi)/subs(d2f,x,xi));
   dx = abs(xii-xi);
   xi = xii;
end

% OUTPUT
x_opt = xii
```

Example 3.7 *Task*: Find a minimum of $f(x) = -\frac{x}{1+x^2}$ using the Newton method. Let the initial estimate be

(a) $x_{(1)} = 0.5$
(b) $x_{(1)} = 2.5$ and the desired accuracy $\varepsilon = 0.01$.

Solution: It is necessary to calculate $f'(x)$ and $f''(x)$ first.

$$f'(x) = \frac{x^2 - 1}{\left(x^2 + 1\right)^2},\ f''(x) = -\frac{2x\left(x^2 - 3\right)}{\left(x^2 + 1\right)^3}$$

Considering (3.45), it can be computed for option (a) that

$$\begin{aligned}
x_{(1)} &= 0.5 \\
x_{(2)} &= 0.5 - \frac{-0.48}{1.408} = 0.8465 \\
x_{(3)} &= 0.8465 - \frac{-1.005}{0.7751} = 0.9724 \\
x_{(4)} &= 0.9724 - \frac{-0.0154}{0.5454} = 0.9989 \\
x_{(5)} &= 0.9989 - \frac{-5.4842 \times 10^{-4}}{0.5016} \approx 1
\end{aligned}$$

Clearly, the sequence of iterated minimum estimates converges to the analytical extremum at $x^* = 1$. It is worth noting that the initial guess $x_{(1)} = 0.5$ lies in the convex area of the $f(x)$, see Fig. 3.2.

Now, look at option (b). By iterating (3.45), starting from $x_{(1)} = 2.5$, one can obtain the following sequence

$$
\begin{aligned}
x_{(1)} &= 2.5 \\
x_{(2)} &= 2.5 - \frac{0.0999}{-0.0426} = 4.8423 \\
x_{(3)} &= 4.8423 - \frac{0.0376}{-0.0136} = 7.6136 \\
x_{(4)} &= 7.6136 - \frac{0.0164}{-0.0041} = 11.6270 \\
x_{(5)} &= 11.6270 - \frac{0.0072}{-0.0012} = 17.5721
\end{aligned}
$$

that clearly diverges to infinity, where a (local) supremum $f(x) \to 0^-$ exists (see Fig. 3.2 again). The reason for this erroneous result has been explained above; see (3.46) and text around.

3.2.1.2 Regula Falsi Method

In principle, the difference between the regula falsi method and the Newton one lies in a numerical approximation of $f''(x)$ and (possibly) $f'(x)$ in (3.45) at some points near the extremum estimate. This concept is usable mainly whenever an analytic formula for $f(x)$ is unavailable, and only some pairs $\left[x_j, f\left(x_j\right)\right]$ are known (e.g., as the results of laboratory measurements).

The simplest approximations can be computed as

$$
\begin{aligned}
f'\left(x_j\right) &\approx \frac{f(x_k) - f\left(x_j\right)}{x_k - x_j} := \tilde{f}'\left(x_j\right) \\
f''\left(x_j\right) &\approx \frac{f'(x_k) - f'\left(x_j\right)}{x_k - x_j} := \tilde{f}''\left(x_j\right)
\end{aligned}
\tag{3.49}
$$

for some sufficiently close points $x_j \neq x_k$. In fact, if $f''\left(x_j\right)$ is approximated using $\tilde{f}'\left(x_j\right)$ and $\tilde{f}'(x_k)$ in (3.49), three different points $x_j \neq x_k \neq x_l$ are needed, since

$$
f'(x_k) \approx \frac{\tilde{f}(x_l) - \tilde{f}(x_k)}{x_l - x_k} := \tilde{f}'(x_k) \tag{3.50}
$$

Hence, two different versions of the algorithm are introduced below. First, a three-initial-points version considers approximating both $f'(x)$ and $f''(x)$ (i.e., $\tilde{f}'\left(x_j\right)$ and

$\tilde{f}'(x_k)$); while, second, a two-initial-points version approximates only $f''(x)$ (i.e., $\tilde{f}''(x)$), by taking $f'(x)$ exact (analytic).

The most significant disadvantage of the former method is that the extremum estimates $(x_j, x_k, x_l = x^1, x^2, x^3)$ must be sufficiently close to the actual extremum and, simultaneously, these estimates must be "dense". It means that if any two of the conditions do not hold, the iterative computations might yield a sequence of further estimates that are "far" from each other or "far" from the factual extremum. This implies that the sequence may diverge (due to an erroneous difference approximation or straying in a convex/concave area of the objective function domain).

The two variants of the method algorithms follow.

Algorithm 3.8a (The regula falsi method—three initial points)

1. Let three initial extremum guesses $x^1_{(1)}, x^2_{(1)}, x^3_{(1)}$ and their corresponding objective function values $f\left(x^1_{(1)}\right), f\left(x^2_{(1)}\right), f\left(x^3_{(1)}\right)$, respectively, be known. Select $\varepsilon > 0$, and set $i = 1$.
2. Compute first derivative approximations (3.49) and (3.50) at $x^1_{(i)}$ and $x^2_{(i)}$ (i.e., $\tilde{f}'\left(x^1_{(i)}\right)$ and $\tilde{f}'\left(x^2_{(i)}\right)$, respectively), using all three extremum estimates and their objective function values.
3. Compute $\tilde{f}''\left(x^1_{(i)}\right)$ using (3.49) with $\tilde{f}'\left(x^1_{(i)}\right)$ and $\tilde{f}'\left(x^2_{(i)}\right)$.
4. Set $x^3_{(i+1)} = x^2_{(i)}, x^2_{(i+1)} = x^1_{(i)}$, and compute the updated $x^1_{(i+1)}$ via (3.45), i.e.

$$x^1_{(i+1)} = x^1_{(i)} - \frac{\tilde{f}'\left(x^1_{(i)}\right)}{\tilde{f}''\left(x^1_{(i)}\right)} \tag{3.51}$$

5. If $\left|x^1_{(i+1)} - x^2_{(i+1)}\right| \geq \varepsilon$, set $i = i + 1$ and go to step 2. Otherwise, stop and return $x^* = x^1_{(i+1)}$.

Algorithm 3.8b (The regula falsi method—two initial points)

1. Let two initial extremum guesses $x^1_{(1)}, x^2_{(1)}$ and their corresponding objective function values $f\left(x^1_{(1)}\right), f\left(x^2_{(1)}\right)$, respectively, be known. Select $\varepsilon > 0$, and set $i = 1$.
2. Calculate $f'\left(x^1_{(i)}\right)$ and $f'\left(x^2_{(i)}\right)$ analytically.
3. Compute $\tilde{f}''\left(x^1_{(i)}\right)$ using (3.49) with $f'\left(x^1_{(i)}\right)$ and $f'\left(x^2_{(i)}\right)$.
4. Set $x^2_{(i+1)} = x^1_{(i)}$, and compute the updated $x^1_{(i+1)}$ via (3.45) (or, equivalently, using (3.51) with $f'\left(x^1_{(i)}\right)$ instead of $\tilde{f}'\left(x^1_{(i)}\right)$).
5. If $\left|x^1_{(i+1)} - x^2_{(i+1)}\right| \geq \varepsilon$, set $i = i + 1$ and go to step 2. Otherwise, stop and return $x^* = x^1_{(i+1)}$.

A possible Matlab code for Algorithm 3.8a:

```
% INPUTS
syms f x
f = input('Enter the objective function: ');
x_init = input('Enter three initial points (extremum guesses) [x1
x2 x3]: ');
e = input('Enter the desired accuracy: ');

% INIT
x1 = x_init(1); x2 = x_init(2); x3 = x_init(3);
df1 = (subs(f,x,x2)-subs(f,x,x1))/(x2-x1);
df2 = (subs(f,x,x3)-subs(f,x,x2))/(x3-x2);
d2f = (df2-df1)/(x2-x1);

% ITERATIONS
while abs(x2-x1) >= e
   x3 = x2;
   x2 = x1;
   x1 = x1 - eval(df1/d2f);
   df1 = (subs(f,x,x2)-subs(f,x,x1))/(x2-x1);
   df2 = (subs(f,x,x3)-subs(f,x,x2))/(x3-x2);
   d2f = (df2-df1)/(x2-x1);
end

% OUTPUT
x_opt = x1
```

A sketch of a Matlab code for Algorithm 3.8b:

```
% INPUTS
syms f x
f = input('Enter the objective function: ');
x_init = input('Enter two initial points (extremum guesses) [x1 x2]:
');
e = input('Enter the desired accuracy: ');

% INIT
x1 = x_init(1); x2 = x_init(2);
df = diff(f,x);
df1 = subs(df,x,x1);
df2 = subs(df,x,x2);
d2f = (df2-df1)/(x2-x1);

% ITERATIONS
while abs(x2-x1) >= e
   x2 = x1;
   x1 = x1 - eval(df1/d2f);
   df1 = subs(df,x,x1);
   df2 = subs(df,x,x2);
   d2f = (df2-df1)/(x2-x1);
end
```

```
% OUTPUT
x_opt = x1
```

The following example demonstrates the effect of initial solution estimates density on the numerical results of both the regula falsi method versions.

Example 3.8 *Task*: Find a minimum of $f(x) = -\frac{x}{1+x^2}$ using the regula falsi method.

Let the initial estimate and the desired accuracy be

(a) $x^1_{(1)} = 0.3,\ x^2_{(1)} = 0.5,\ x^3_{(1)} = 0.7, \varepsilon = 0.01$
(b) $x^1_{(1)} = 0.89,\ x^2_{(1)} = 0.9,\ x^3_{(1)} = 0.91, \varepsilon = 0.001.$

Solution: Consider option (a) with Algorithm 3.8a first. The following sequence of optimum estimate triples is obtained.

$$x^1_{(1)} = 0.3,\ x^2_{(1)} = 0.5,\ x^3_{(1)} = 0.7$$

$$\begin{aligned} x^1_{(2)} &= 0.3 - \frac{\frac{-0.4698-(-0.4)}{0.7-0.5} - \frac{-0.4-(-0.2752)}{0.5-0.3}}{0.5 - 0.3} \\ &= 0.754,\ x^2_{(2)} = 0.3,\ x^3_{(2)} = 0.5 \end{aligned}$$

$$x^1_{(3)} = 0.754 - \frac{\frac{-0.4-(-0.2752)}{0.5-0.3} - \frac{-0.2752-(-0.4807)}{0.3-0.754}}{0.3 - 0.754} = 1.954,$$
$$x^2_{(3)} = 0.754,\ x^3_{(3)} = 0.3$$

$$x^1_{(4)} = 1.954 - \frac{\frac{-0.2752-(-0.4807)}{0.3-0.754} - \frac{-0.4807-(-0.4056)}{0.754-1.954}}{0.754 - 1.954} = 1.8081,$$
$$x^2_{(4)} = 1.954,\ x^3_{(4)} = 0.754$$

$$x^1_{(5)} = 1.8081 - \frac{\frac{-0.4807-(-0.4056)}{0.754-1.954} - \frac{-0.4056-(-0.4235)}{1.954-1.8081}}{1.954 - 1.8081} = 2.1049,$$
$$x^2_{(5)} = 1.8081,\ x^3_{(5)} = 1.954$$

Apparently, the sequence of minimum estimates obtained by iterating (3.51) does not converge. The values are swept very soon, mainly due to inaccurate approximations $\tilde{f}'\left(x^1_{(i)}\right)$ and $\tilde{f}'\left(x^2_{(i)}\right)$ (already for $i = 1, 2$).

Second, assume Algorithm 3.8b. The analytic formula for $f'(x)$ has already been given in Example 3.7. Using the formula, one can compute the following iterative sequence of the minimum estimates.

$$x^1_{(1)} = 0.3,\ x^2_{(1)} = 0.5,\ x^3_{(1)} = 0.7$$

$$x^1_{(2)} = 0.3 - \frac{-0.48 - (-0.7659)}{0.5 - 0.3} = 0.8357,$$
$$x^2_{(2)} = 0.3,\ x^3_{(2)} = 0.5$$
$$x^1_{(3)} = 0.8357 - \frac{-0.7659 - (-0.1045)}{0.3 - 0.8357} = 0.9204,$$
$$x^2_{(3)} = 0.8357,\ x^3_{(3)} = 0.3$$
$$x^1_{(4)} = 0.9204 - \frac{-0.1045 - (-0.0448)}{0.8357 - 0.9204} = 0.9839,$$
$$x^2_{(4)} = 0.9204,\ x^3_{(4)} = 0.8357$$
$$x^1_{(5)} = 0.9839 - \frac{-0.448 - (-0.0082)}{0.9204 - 0.9839} = 0.9982,$$
$$x^2_{(5)} = 0.9839,\ x^3_{(5)} = 0.9204$$

The desired accuracy is reached after the 6th iteration, since it can be computed that $x^1_{(6)} = 0.99996$ and $|0.9999 - 0.9982| = 0.0017 < 0.01$. In this example, the sequence of minimum estimates converges despite the fact that the approximation $\tilde{f}''\left(x^1_{(i)}\right)$ is not accurate.

Now, let us have a look at option (b) of the example assignment. The initial guesses are not so sparse as it was for (a). Intuitively, the derivative approximation might be more accurate.

Considering Algorithm 3.8a with three initial estimates, the following iteration steps can be computed

$$x^1_{(1)} = 0.89,\ x^2_{(1)} = 0.9,\ x^3_{(1)} = 0.91$$

$$x^1_{(2)} = 0.89 - \frac{\frac{-0.4978-(-0.4972)}{0.91-0.9} - \frac{-0.4972-(-0.4966)}{0.9-0.89}}{0.9 - 0.89} = 0.9823,$$
$$x^2_{(2)} = 0.89,\ x^3_{(2)} = 0.9$$
$$x^1_{(3)} = 0.9823 - \frac{\frac{-0.4972-(-0.4966)}{0.9-0.89} - \frac{-0.4966-(-0.4999)}{0.89-0.9823}}{0.3 - 0.754} = 1.1109,$$
$$x^2_{(3)} = 0.9823,\ x^3_{(3)} = 0.89$$
$$x^1_{(4)} = 1.1109 - \frac{\frac{-0.4966-(-0.4999)}{0.89-0.9823} - \frac{-0.4999-(-0.4972)}{0.9823-1.1109}}{0.9823 - 1.1109} = 1.0636,$$
$$x^2_{(4)} = 1.1109,\ x^3_{(4)} = 0.9823$$
$$x^1_{(5)} = 1.0636 - \frac{\frac{-0.4999-(-0.4972)}{0.9823-1.1108} - \frac{-0.4972-(-0.4991)}{1.1108-1.0636}}{1.1108 - 1.0636} = 1.1675,$$
$$x^2_{(5)} = 1.0636,\ x^3_{(5)} = 1.1108$$

Although it seems that the iterative sequence of minimum estimates $x^1_{(i)}$ keeps close to the factual extremum $x^* = 1$, next iteration steps do not confirm this expectation:

$$x^1_{(6)} = 0.6625,\ x^1_{(7)} = 0.956,\ x^1_{(8)} = 0.3686,\ x^1_{(9)} = 1.4305,$$
$$x^1_{(10)} = 2.33,\ x^1_{(11)} = 1.911,\ x^1_{(12)} = -11.2272, \ldots$$

Hence, it can be observed that despite a relatively dense triplet of initial extremum guesses, the "spacing" between $x^1_{(i)}$ becomes higher in further iteration steps. These values then yield less accurate derivative approximations, which eventually results in the solution divergence. It can be concluded that Algorithm 3.8a is usable only in case of sufficiently dense initial estimates and using only a limited number of iteration steps (e.g., until the solution difference between consecutive iterations exceeds a chosen threshold).

Contrariwise, significantly different numerical outputs are obtained in option (b) when using Algorithm 3.8b.

$$x^1_{(1)} = 0.89,\ x^2_{(1)} = 0.9,\ x^3_{(1)} = 0.91$$

$$x^1_{(2)} = 0.89 - \frac{-0.058 - (-0.0647)}{0.9 - 0.89} = 0.9861,\ x^2_{(2)} = 0.89,\ x^3_{(2)} = 0.9$$

$$x^1_{(3)} = 0.9861 - \frac{-0.0647 - (-0.0071)}{0.89 - 0.9861} = 0.9979,\ x^2_{(3)} = 0.9861,\ x^3_{(3)} = 0.89$$

$$x^1_{(4)} = 0.9979 - \frac{-0.0071 - (-0.001)}{0.9861 - 0.9979} = 0.9999,$$
$$x^2_{(4)} = 0.9979,\ x^3_{(4)} = 0.9861$$

$$x^1_{(5)} = 0.9999 - \frac{-0.001 - \left(-2.1 \times 10^{-5}\right)}{0.9979 - 0.9999} \approx 1,$$
$$x^2_{(5)} = 0.9999,\ x^3_{(5)} = 0.9979$$

The obtained result is comparable to that obtained by the Newton method in Example 3.7.

3.2.2 *Multi-dimensional Unconstrained Problem*

An unconstrained multi-dimensional optimization problem (2.15) is assumed in this section. Besides a generalized Newton method and some of its extensions (the so-called quasi-Newton methods), the very basic gradient-based steepest-descent method is described.

3.2.2.1 Newton Method

The method generalizes the one-dimensional case introduced in Sect. 3.2.1.1. Its iterative rule can be derived as follows.

Let $f(\mathbf{x})$ have first and second partial derivatives at fixed point $\mathbf{x}_{(i)} \in \mathbb{R}^n$ in the neighborhood of the function extreme $\mathbf{x}^*$. Then, the corresponding Taylor series reads

$$\begin{aligned} f(\mathbf{x}) = & f\left(\mathbf{x}_{(i)}\right) + \left[\operatorname{grad} f\left(\mathbf{x}_{(i)}\right)\right]^T\left(\mathbf{x} - \mathbf{x}_{(i)}\right) \\ & + \frac{1}{2}\left(\mathbf{x} - \mathbf{x}_{(i)}\right)^T \mathbf{H}\left(f\left(\mathbf{x}_{(i)}\right)\right)\left(\mathbf{x} - \mathbf{x}_{(i)}\right) + R\left(\left\|\mathbf{x} - \mathbf{x}_{(i)}\right\|\right) \end{aligned} \tag{3.52}$$

The necessary extremum condition $\operatorname{grad} f(\mathbf{x})|_{\mathbf{x}=\mathbf{x}^*} \overset{!}{=} 0$ (see (2.13) and Theorem 2. 2) imply from (3.52) that

$$\begin{aligned} \operatorname{grad} f\left(\mathbf{x}^*\right) = & \operatorname{grad} f\left(\mathbf{x}_{(i)}\right) + \mathbf{H}\left(f\left(\mathbf{x}_{(i)}\right)\right)\left(\mathbf{x}^* - \mathbf{x}_{(i)}\right) \\ & + R\left(\left\|\mathbf{x}^* - \mathbf{x}_{(i)}\right\|\right) \overset{!}{=} \mathbf{0}_n \end{aligned} \tag{3.53}$$

If the Hessian $\mathbf{H}\left(f\left(\mathbf{x}_{(i)}\right)\right)$ is regular and remainder $R\left(\left\|\mathbf{x}^* - \mathbf{x}_{(i)}\right\|\right)$ is neglected, then condition (3.53) implies

$$\tilde{\mathbf{x}}^* = \mathbf{x}_{(i)} - \left[\mathbf{H}\left(f\left(\mathbf{x}_{(i)}\right)\right)\right]^{-1} \operatorname{grad} f\left(\mathbf{x}_{(i)}\right) \tag{3.54}$$

where $\tilde{\mathbf{x}}^*$ expresses an extremum estimate. Considering iterative updates of this estimate, the eventual multi-dimensional Newton method rule reads

$$\mathbf{x}_{(i+1)} = \mathbf{x}_{(i)} - \left[\mathbf{H}\left(f\left(\mathbf{x}_{(i)}\right)\right)\right]^{-1} \operatorname{grad} f\left(\mathbf{x}_{(i)}\right) \tag{3.55}$$

However, the method suffers from the identical drawback as described in the one-dimensional case. In particular, when searching for a *minimum*, the quadratic form with the Hessian must be *positive definite*, and vice versa (see Theorem 2.2). Hence, analogously to (3.48), the following trick can be made. Whenever the quadratic form with $\mathbf{H}\left(f\left(\mathbf{x}_{(i)}\right)\right)$ is not positive definite, one can make the following update

$$\mathbf{M}\left(\mathbf{x}_{(i)}, \delta_{(i)}\right) = \mathbf{H}\left(f\left(\mathbf{x}_{(i)}\right)\right) + \delta_{(i)} \mathbf{I}_n \tag{3.56}$$

where $\delta_{(i)} > 0$ is selected so that $\mathbf{M}\left(\mathbf{x}_{(i)}, \delta_{(i)}\right)$ is positive definite, and $\mathbf{I}_n$ means the $n \times n$ unit matrix. Then, $\mathbf{H}\left(f\left(\mathbf{x}_{(i)}\right)\right)$ is substituted by $\mathbf{M}\left(\mathbf{x}_{(i)}, \delta_{(i)}\right)$ in (3.55).

Another computational disadvantage of the Newton method is the necessity of the knowledge of the Hessian at each iteration step, which can cause a burden, mainly for large-dimension problems.

Let the termination condition be

$$\left\|\mathbf{x}_{(i+1)} - \mathbf{x}_{(i)}\right\| < \varepsilon \tag{3.57}$$

The algorithm can be summarized as follows:

Algorithm 3.9 (The multi-dimensional Newton method)

1. Let objective function $f(\mathbf{x})$ (having partial derivatives up to the order of two) and the initial extremum estimate $\mathbf{x}_{(1)} \in \mathbb{R}^n$ be given. Select $\varepsilon > 0$, and set $i = 1$.
2. Calculate grad$f(\mathbf{x})$ and $\mathbf{H}(f(\mathbf{x}))$.
3. Compute $\mathbf{x}_{(i+1)}$ using (3.55).
4. If (3.57) does not hold, set $i = i + 1$ and go to step 3. Otherwise, stop and return $\mathbf{x}^* = \mathbf{x}_{(i+1)}$.

A Matlab code for the two-variable case:

```
% INPUTS
syms f x1 x2
f = input('Enter the objective function: ');
x_init = input('Enter the initial point (extremum guess) [x1 x2]:
');
e = input('Enter the desired accuracy: ');

% INIT
n = 2;
grad_f = gradient(f);
H_f= hessian(f);
xi = x_init';
xii = x_init' + sqrt(2)*e*ones(n,1);
dx = norm((xii-xi));

% ITERATIONS
while dx >= e
  xii = xi - eval(subs(H_f,[x1;x2],xi)\subs(grad_f,[x1;x2],xi));
  dx = norm((xii-xi));
  xi = xii;
end
% OUTPUT
x_opt = xii
```

The following simple example demonstrates that the Newton method reaches the exact extremum in a single step for quadratic (convex or concave) functions. Indeed, the reminder $R(\cdot)$ in (3.52) is zero in these cases, and hence, expression (3.54) is exact (i.e., the analytic solution).

Example 3.9 *Task*: Find a minimum of $f(\mathbf{x}) = f(x_1, x_2) = 3x_1^2 - 2x_1x_2 + x_2^2 + 4x_1 + 3x_2 + 1$ on $\mathbb{R}^2$ using the Newton method. Let the initial estimate be $\mathbf{x}_{(1)} = [0, 0]^T$.

Solution: The gradient and Hessian of the objective function can be calculated analytically:

$$\text{grad}f(\mathbf{x}) = [6x_1 - 2x_2 + 4, 2x_2 - 2x_1 + 3]^T$$

$$\mathbf{H}(f(\mathbf{x})) = \begin{bmatrix} 6 & -2 \\ -2 & 2 \end{bmatrix}$$

It can be noticed that the Hessian is constant and moreover, positive definite for all $\mathbf{x} \in \mathbb{R}^2$ (see Proposition 2.4). Therefore, the function is convex as per Proposition 2.5.

The Hessian inversion equals

$$\mathbf{H}^{-1}(f(\mathbf{x})) = \frac{1}{4}\begin{bmatrix} 1 & 1 \\ 1 & 3 \end{bmatrix}$$

and the first iteration step reads

$$\begin{aligned} \mathbf{x}_{(2)} &= \mathbf{x}_{(1)} - \mathbf{H}^{-1}\left(f\left(\mathbf{x}_{(1)}\right)\right)\mathrm{grad}\,f\left(\mathbf{x}_{(1)}\right) \\ &= \begin{bmatrix} 0 \\ 0 \end{bmatrix} - \frac{1}{4}\begin{bmatrix} 1 & 1 \\ 1 & 3 \end{bmatrix}\begin{bmatrix} 4 \\ 3 \end{bmatrix} = \begin{bmatrix} -1.75 \\ -3.25 \end{bmatrix} \end{aligned}$$

that represents the exact local and global minimum.

3.2.2.2 Steepest Descent Method

This method represents the fundamental gradient-based technique. It purely adopts the general rule (3.42) with a constant or optimized $\lambda_{(i)}$. The following proposition extends Proposition 3.1.

Proposition 3.2 *If* $\mathbf{d} \in \mathbb{R}^n$ *is the direction of* $f(\mathbf{x})$ *value increase in the neighborhood of some (at which the function is differentiable)* $\mathbf{x}_{(i)} \in \mathbb{R}^n$*, i.e.,* $f\left(\mathbf{x}_{(i)} + \lambda\mathbf{d}\right) > f\left(\mathbf{x}_{(i)}\right)$ *for* $\lambda \in [0, \lambda_{\max}]$ *(or* $f\left(\mathbf{x}_{(i)} + \lambda\mathbf{d}\right) < f\left(\mathbf{x}_{(i)}\right)$ *for* $\lambda \in [-\lambda_{\max}, 0]$*), then it can be shown that the steepest ascent direction is*

$$\mathbf{d} = \mathrm{grad} f\left(\mathbf{x}_{(i)}\right) \tag{3.58}$$

If $\lambda_{(i)}$ is optimized at each iteration step, the given multi-dimensional problem is transformed into a one-dimensional subproblem that is solved analytically (similarly to the cyclic parameter sweeping method described in Sect. 3.1.2.3). In this case, the geometric interpretation can roughly be as follows.

Consider, for instance, the searching for a minimum of $f(\mathbf{x})$. The variable space is searched in the gradient direction from the current point as long as the function value of $f(\mathbf{x})$ decreases. That is, a jump to the nearest one-dimensional local minimum along the gradient is done, see Fig. 3.15b. Let this technique be called **long-step**.

In contrast, if $\lambda_{(i)} = \lambda$ is constant at each step, the gradient vector is lengthened by the same multiplier (see Fig. 3.15a). It represents the **short-step** steepest descent method.

Termination condition (3.57) can be adopted. However, the iterative procedure should also be stopped whenever $\mathrm{grad} f\left(\mathbf{x}_{(i)}\right) \approx \mathbf{0}_{n\times 1}$, as $\lambda_{(i)}(\mathbf{x})$ may then vanish when computing.

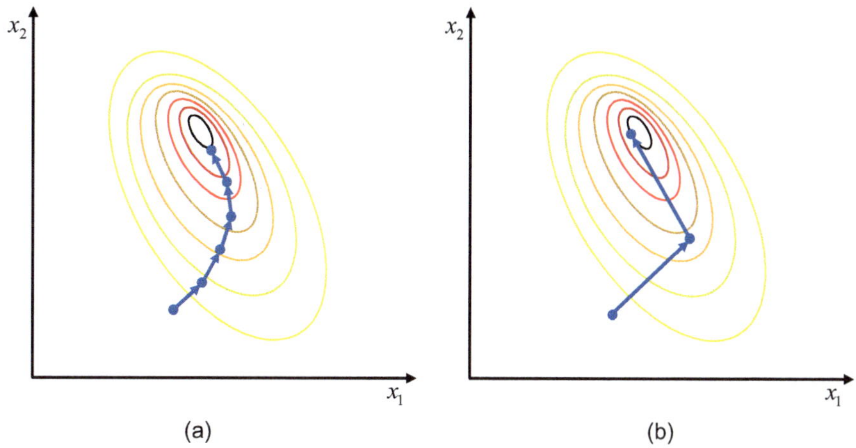

Fig. 3.15 A geometric interpretation of the short-step (**a**) and long-step (**b**) steepest descent (gradient) methods with $\lambda_{(i)} = \lambda = \text{const.}$ and $\lambda_{(i)}$ optimized, respectively

Algorithm 3.10a describes the short-step technique, while the long-step one is presented in Algorithm 3.10b. Each of the algorithms is followed by a sketch of the Matlab code.

Algorithm 3.10a (The short-step steepest descent method)

1. Let objective function $f(\mathbf{x})$ (the gradient of which exists) and the initial extremum estimate $\mathbf{x}_{(1)} \in \mathbb{R}^n$ be given. Select the desired accuracy $\varepsilon > 0$, gradient norm threshold $\varepsilon \gg e > 0$, and step length λ. Set $i = 1$.
2. Calculate $\text{grad} f(\mathbf{x})$.
3. Compute $\text{grad} f\left(\mathbf{x}_{(i)}\right)$. If $\left\|\text{grad} f\left(\mathbf{x}_{(i)}\right)\right\| < e$, set $i = i - 1$ and go to step 6; else, go to step 4.
4. Compute $\mathbf{x}_{(i+1)}$ using (3.42) with the selected constant λ.
5. If (3.57) holds, go to step 6. Otherwise, set $i = i + 1$ and go to step 3.
6. Stop and return $\mathbf{x}^* = \mathbf{x}_{(i+1)}$.

```
% INPUTS
syms f x1 x2
f = input('Enter the objective function: ');
x_init = input('Enter the initial point [x1 x2]: ');
lam = input('Enter the step length (lambda): ');
e = input('Enter the desired accuracy: ');

% INIT
n = 2;
grad_f = gradient(f);
x_prev = x_init';
x = x_init';
dx = inf;
```

```
% ITERATIONS
while dx >= e
   grad_fx = subs(grad_f,[x1;x2],x);
   if eval(norm(grad_fx)<eps)
      break
   end
   x = x+lam*grad_fx;
   dx = norm(x-x_prev);
   x_prev = x;
end

% OUTPUT
x
```

Algorithm 3.10b (The long-step steepest descent method)

1. Let objective function $f(\mathbf{x})$ (the gradient of which exists) and the initial extremum estimate $\mathbf{x}_{(1)} \in \mathbb{R}^n$ be given. Select the desired accuracy $\varepsilon > 0$ and gradient norm threshold $\varepsilon \gg e > 0$. Set $i = 1$.
2. Calculate $\operatorname{grad} f(\mathbf{x})$.
3. Compute $\operatorname{grad} f\left(\mathbf{x}_{(i)}\right)$. If $\left\|\operatorname{grad} f\left(\mathbf{x}_{(i)}\right)\right\| < e$, set $i = i - 1$ and go to step 8; else, go to step 4.
4. Construct the following function

$$p\left(\lambda_{(i)}\right) := f\left(\mathbf{x}_{(i)} + \lambda_{(i)} \operatorname{grad} f\left(\mathbf{x}_{(i)}\right)\right) \tag{3.59}$$

5. Solve (analytically) the one-dimensional optimization subproblem

$$\lambda_{(i)}^{*} = \arg \operatorname{extr} p\left(\lambda_{(i)}\right) \tag{3.60}$$

6. Compute $\mathbf{x}_{(i+1)}$ using (3.42) with the optimized $\lambda_{(i)}^{*}$.
7. If (3.57) holds, go to step 8. Otherwise, set $i = i + 1$ and go to step 3.
8. Stop and return $\mathbf{x}^{*} = \mathbf{x}_{(i+1)}$.

```
% INPUTS
syms f x1 x2 lam
f = input('Enter the objective function: ');
x_init = input('Enter the initial point [x1 x2]: ');
e = input('Enter the desired accuracy: ');

% INIT
n = 2;
grad_f = gradient(f);
x_prev = x_init';
x = x_init';
dx = inf;


% ITERATIONS
```

```
while dx >= e
   grad_fx = subs(grad_f,[x1;x2],x);
   if eval(norm(grad_fx)<eps)
      break
   end
   p = subs(f,[x1;x2],x+lam*grad_fx);
   % ONE-DIMENSIONAL OPTIMIZATION
   dp = diff(p,lam);
   lam0 = solve(dp==0);
   n_lam0 = length(lam0);
   lam_min=inf;
   for j = 1:1:n_lam0
      l = 2;
      while eval(subs(diff(dp,lam),lam,lam0(j))==0)
         dp = diff(dp,lam);
         l = l+1;
      end
     if (rem(l,2)==0) && eval(subs(diff(dp,lam),lam,lam0(j))>0) &&
(lam0(j)<0)
         if lam0(j) < lam_min
            lam_min = lam0(j); % A BETTER SOLUTION FOUND
         end
      end
   end
   x = x+lam_min*grad_fx;
   dx = norm(x-x_prev);
   x_prev = x;
end

% OUTPUT
x
```

Note that the gradient norm in step 3 of Algorithms 3.10a and 3.10b can be normalized. Recall that the optimized $\lambda^*_{(i)}$ must be positive when searching for an objective function maximum and vice versa (for a minimum).

A disadvantage of the method can be a slow solution convergence near the extremum. It is called the *zigzag* solution path.

Example 3.10 *Task*: Find a minimum of $f(\mathbf{x}) = f(x_1, x_2) = 3x_1^2 - 2x_1x_2 + x_2^2 + 4x_1 + 3x_2 + 1$ on $\mathbb{R}^2$ using

(a) the short-step steepest descent method with $\lambda = -0.3$
(b) the long-step steepest descent method

starting from the initial estimate $\mathbf{x}_{(1)} = [0, 0]^T$. Let the given accuracy be $\varepsilon = 0.05$.

Solution: The gradient of $f(\mathbf{x})$ has been calculated in Example 3.9.

Let task (a) be considered. The first iterative step reads

$$\mathbf{x}_{(2)} = \mathbf{x}_{(1)} + \lambda \text{grad} f\left(\mathbf{x}_{(1)}\right) = [0, 0]^T - 0.3[4, 3]^T = [-1.2, -0.9]^T$$

From this point, the following solution estimate is searched in the opposite direction to $\text{grad} f(\mathbf{x}_{(2)})$ multiplied by 0.3, i.e.,

$$\mathbf{x}_{(3)} = \mathbf{x}_{(2)} + \lambda \text{grad} f(\mathbf{x}_{(2)}) = [-1.2, -0.9]^T - 0.3[-1.4, -3.6]^T$$
$$= [-0.78, -1.98]^T$$

Results of the further iteration steps are summarized in Table 3.7.

After the 10th iteration, the obtained optimum estimation is $\mathbf{x}_{(11)} = [-1.1779, -3.4346]^T$ with $f(\mathbf{x}_{(11)}) = -6.1479$. From the last few iteration steps in Table 3.7; it is clear that the minimum estimate is close to the actual extreme $\mathbf{x}^* = [-1.75, -3.25]^T, f(\mathbf{x}^*) = -7.375$. However, there is a "zigzag" movement of the solution in the variable space and a deterioration of the estimate (increase of functional values), starting from $i = 6$, see Fig. 3.16. Further computations can show that the solution diverges to infinity. This is due to the relatively high values of the

Table 3.7 Short-step steepest descent method—Example 3.10

i	$\mathbf{x}_{(i)}$	$\text{grad} f(\mathbf{x}_{(i)})$	$\lVert \mathbf{x}_{(i+1)} - \mathbf{x}_{(i)} \rVert$	$f(\mathbf{x}_{(i)})$
1	$\begin{bmatrix} 0 \\ 0 \end{bmatrix}$	$\begin{bmatrix} 4 \\ 3 \end{bmatrix}$	–	1
2	$\begin{bmatrix} -1.2 \\ -0.9 \end{bmatrix}$	$\begin{bmatrix} -1.4 \\ -3.6 \end{bmatrix}$	1.5	– 3.53
3	$\begin{bmatrix} -0.78 \\ -1.98 \end{bmatrix}$	$\begin{bmatrix} 3.28 \\ 0.6 \end{bmatrix}$	1.1588	– 5.4032
4	$\begin{bmatrix} -1.764 \\ -2.16 \end{bmatrix}$	$\begin{bmatrix} -2.264 \\ 2.208 \end{bmatrix}$	1.0003	– 6.1558
5	$\begin{bmatrix} -1.0848 \\ -2.8224 \end{bmatrix}$	$\begin{bmatrix} 3.136 \\ -0.4752 \end{bmatrix}$	0.9487	– 6.4336
6	$\begin{bmatrix} -2.0256 \\ -2.6798 \end{bmatrix}$	$\begin{bmatrix} -2.7939 \\ 1.6915 \end{bmatrix}$	0.9515	– 6.5078
7	$\begin{bmatrix} -1.1874 \\ -3.1875 \end{bmatrix}$	$\begin{bmatrix} 3.25 \\ -0.9997 \end{bmatrix}$	0.9798	– 6.4921
8	$\begin{bmatrix} -2.1624 \\ -2.8874 \end{bmatrix}$	$\begin{bmatrix} -3.1996 \\ 1.55 \end{bmatrix}$	1.0201	– 6.4342
9	$\begin{bmatrix} -1.2025 \\ -3.3524 \end{bmatrix}$	$\begin{bmatrix} 3.4897 \\ -1.2998 \end{bmatrix}$	1.0667	– 6.3532
10	$\begin{bmatrix} -2.2494 \\ -2.9625 \end{bmatrix}$	$\begin{bmatrix} -3.5716 \\ 1.5739 \end{bmatrix}$	1.1173	– 6.2569

Fig. 3.16 The "zigzag" solution near the optimum when using the short-step steepest descent method—Example 3.10

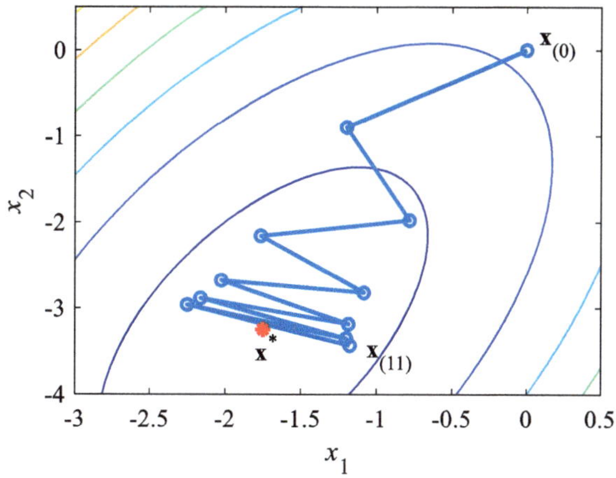

gradient entries. The prolonged opposite gradient always jumps over the minimum, and this "sweeping" graduates.

The solution can be improved by changing the step length λ or terminating the iterations whenever the function values start to increase.

Now, let us have a look at task (b). In contrast to the short-step method, the step length is properly reduced, so that a solution estimate in the particular iteration does not "jump over" the best solution in the opposite gradient direction. As the given cost function is simple enough, the analytic solution to $\lambda^*_{(i)}\left(x_{1,(i)}, x_{2,(i)}\right)$ can be calculated first, which enables the expression of the optimal step length for each iteration explicitly.

Hence, the following one-dimensional subtask is solved

$$\lambda^* = \arg\min_{\lambda<0} p(\lambda),\ p(\lambda) = f(\mathbf{x} + \lambda \operatorname{grad} f(\mathbf{x}))$$

see (3.59) and (3.60). Function $p\left(\lambda_{(i)}\right)$ reads

$$\begin{aligned}
p\left(\lambda_{(i)}\right) &= f(\mathbf{x})|_{\mathbf{x}=\mathbf{x}_{(i)}+\lambda_{(i)}\operatorname{grad} f\left(\mathbf{x}_{(i)}\right)} \\
&= 3\left(x_{1,(i)} + \lambda_{(i)}\left(6x_{1,(i)} + 4 - 2x_{2,(i)}\right)\right)^2 \\
&\quad - 2\left(x_{1,(i)} + \lambda_{(i)}\left(6x_{1,(i)} + 4 - 2x_{2,(i)}\right)\right)\left(x_{2,(i)} + \lambda_{(i)}\left(2x_{2,(i)} + 3 - 2x_{1,(i)}\right)\right) \\
&\quad + \left(x_{2,(i)} + \lambda_{(i)}\left(2x_{2,(i)} + 3 - 2x_{1,(i)}\right)\right)^2 \\
&\quad + 4\left(x_{1,(i)} + \lambda_{(i)}\left(6x_{1,(i)} + 4 - 2x_{2,(i)}\right)\right) \\
&\quad + 3\left(x_{2,(i)} + \lambda_{(i)}\left(2x_{2,(i)} + 3 - 2x_{1,(i)}\right)\right) + 1
\end{aligned}$$

As the function is quadratic, the necessary and sufficient optimum condition is $p'(\lambda) = 0$, see Theorem 2.1. After some computation, one gets

$$\lambda_{(i)}^* = 0.5\frac{-40x_{1,(i)}^2 - 8x_{2,(i)}^2 - 36x_{1,(i)} + 4x_{2,(i)} + 32x_{1,(i)}x_{2,(i)} - 25}{136x_{1,(i)}^2 + 24x_{2,(i)}^2 + 112x_{1,(i)} - 40x_{2,(i)} - 112x_{1,(i)}x_{2,(i)} + 33}$$

Hence, it can be computed that $\lambda_{(1)}^* = -\frac{22}{66}$ for $\mathbf{x}_{(1)} = [0, 0]^T$; therefore

$$\begin{aligned}\mathbf{x}_{(2)} &= \mathbf{x}_{(1)} + \lambda_{(1)}^* \mathrm{grad} f\left(\mathbf{x}_{(1)}\right) = [0, 0]^T - \frac{25}{66}[4, 3]^T \\ &= [-50/33, -25/22]^T \\ &= [-1.5152, -1.1364]^T\end{aligned}$$

with $f\left(\mathbf{x}_{(2)}\right) = -3.7349$. In the next iteration step, an updated optimal step length is computed $\lambda_{(2)}^*\left(\mathbf{x}_{(2)}\right) = -0.1866$, giving rise to

$$\begin{aligned}\mathbf{x}_{(3)} &= \mathbf{x}_{(2)} + \lambda_{(2)}^* \mathrm{grad} f\left(\mathbf{x}_{(2)}\right) \\ &= [-1.5152, -1.1364]^T - 0.1866[-2.8182, 3.7576]^T \\ &= [-50/33, -25/22]^T \\ &= [-0.9894, -1.8374]^T\end{aligned}$$

with $f\left(\mathbf{x}_{(3)}\right) = -5.7928$. Results for the following iteration steps are summarized in Table 3.8.

The following 11th iteration eventually yields $\mathbf{x}_{(11)} = [-1.7228, -3.1996]^T$, $f\left(\mathbf{x}_{(11)}\right) = -7.373$, $\left\|\mathbf{x}_{(11)} - \mathbf{x}_{(10)}\right\| = 0.0313$.

Convergence is apparently faster than for the short-step steepest-descent method, even than for the Nelder–Mead method (which is due to the fact that the objective function is "nice", i.e., quadratic), see Example 3.5. Nevertheless, it is surprisingly slower than the Gauss–Seidel method (see Example 3.6), where the cuts (sections) in the plane of variables are always made for each variable separately instead of the gradient direction.

3.2.2.3 Quasi-Newton Methods

These methods remove the disadvantage of the Newton method, which is that it is necessary to calculate the Hessian in each iteration step. Especially in the case of numerical estimation of the Hessian, this can be a demanding calculation.

Consider Formula (3.53). In principle, an iterative sequence $\left\{\mathbf{S}_{(i)}\right\}_{i=1}^N$ satisfying

$$\mathrm{grad} f\left(\mathbf{x}^*\right) = \mathrm{grad} f\left(\mathbf{x}_{(i)}\right) + \mathbf{S}_{(i)}\left(\mathbf{x}^* - \mathbf{x}_{(i)}\right) \tag{3.61}$$

is attempted to be searched, where $\mathbf{S}_{(1)} = \mathbf{I}_n$ (for a minimum) and $\mathbf{x}^* \in \mathbb{R}^n$ is the minimum point. Note that an alternative principle reads $\mathrm{grad} f\left(\mathbf{x}^*\right) = \mathrm{grad} f\left(\mathbf{x}_{(i)}\right) + \lambda_{(i)}\left(\mathbf{x}^* - \mathbf{x}_{(i)}\right)$, $\lambda_{(i)} \in \mathbb{R}$ with $\lambda_{(1)} = 1$.

Table 3.8 Long-step steepest descent method—Example 3.10

i	$\mathbf{x}_{(i)}$	$\text{grad} f(\mathbf{x}_{(i)})$	$\lambda^*_{(i)}$	$\lVert\mathbf{x}_{(i+1)} - \mathbf{x}_{(i)}\rVert$	$f(\mathbf{x}_{(i)})$
1	$\begin{bmatrix} 0 \\ 0 \end{bmatrix}$	$\begin{bmatrix} 4 \\ 3 \end{bmatrix}$	−0.3788	–	1
2	$\begin{bmatrix} -1.5152 \\ -1.1364 \end{bmatrix}$	$\begin{bmatrix} -2.8182 \\ 3.7576 \end{bmatrix}$	−0.1886	1.8939	−3.7349
3	$\begin{bmatrix} -0.9894 \\ -1.8374 \end{bmatrix}$	$\begin{bmatrix} 1.7986 \\ 1.3039 \end{bmatrix}$	−0.3788	0.8763	−5.7928
4	$\begin{bmatrix} -1.6479 \\ -2.3313 \end{bmatrix}$	$\begin{bmatrix} -1.2249 \\ 1.6332 \end{bmatrix}$	−0.1886	0.8332	−6.6873
5	$\begin{bmatrix} -1.4194 \\ -2.636 \end{bmatrix}$	$\begin{bmatrix} 0.7557 \\ 0.5667 \end{bmatrix}$	−0.3788	0.3809	−7.0761
6	$\begin{bmatrix} -1.7056 \\ -2.8507 \end{bmatrix}$	$\begin{bmatrix} -0.5324 \\ 0.7099 \end{bmatrix}$	−0.1886	0.3578	−7.2451
7	$\begin{bmatrix} -1.6063 \\ -2.9831 \end{bmatrix}$	$\begin{bmatrix} 0.3285 \\ 0.2463 \end{bmatrix}$	−0.3788	0.1655	−7.3185
8	$\begin{bmatrix} -1.7307 \\ -3.0764 \end{bmatrix}$	$\begin{bmatrix} -0.2314 \\ 0.3086 \end{bmatrix}$	−0.1886	0.1555	−7.3505
9	$\begin{bmatrix} -1.6875 \\ -3.134 \end{bmatrix}$	$\begin{bmatrix} 0.1428 \\ 0.1071 \end{bmatrix}$	−0.3788	0.072	−7.3643
10	$\begin{bmatrix} -1.7416 \\ -3.1746 \end{bmatrix}$	$\begin{bmatrix} -0.1006 \\ 0.1341 \end{bmatrix}$	−0.1886	0.0676	−7.3704

The core of many quasi-Newton methods is based on applying the long-step steepest descent method (see Sect. 3.2.2.2) to a general quadratic objective (cost function) function

$$F(\mathbf{x}) = (\mathbf{x} - \mathbf{x}^*)^T \mathbf{C}(\mathbf{x} - \mathbf{x}^*) = \mathbf{c}^T \mathbf{x} + \mathbf{x}^T \tilde{\mathbf{C}} \mathbf{x} \tag{3.62}$$

with $\mathbf{x}, \mathbf{c} \in \mathbb{R}^n$, $\mathbf{C}$, $\tilde{\mathbf{C}} \in \mathbb{R}^{n \times n}$ are positive definite. Note that the corresponding gradient and Hessian read

$$\begin{aligned} \text{grad}\, F(\mathbf{x}) = 2\mathbf{C}(\mathbf{x} - \mathbf{x}^*) = -2\mathbf{C}(\mathbf{x}^* - \mathbf{x}) = \mathbf{c} + 2\tilde{\mathbf{C}}\mathbf{x} \\ \mathbf{H}(F(\mathbf{x})) = 2\mathbf{C} = 2\tilde{\mathbf{C}} \end{aligned} \tag{3.63}$$

The first equality in (3.63) can also be expressed as $\mathbf{0} = \text{grad}\,F(\mathbf{x}) + 2\mathbf{C}(\mathbf{x}^* - \mathbf{x})$. Taking into account that $\text{grad}\,F(\mathbf{x}^*) = 0$, the comparison of (3.61) and (3.63) implies that the Hessian satisfies the condition for $\mathbf{S}$ exactly.

Now, let the long-step steepest descent principle (3.59) be used to (3.62):

$$\begin{aligned} p(\lambda_{(i)}) &= \left(\mathbf{x}_{(i)} + \lambda_{(i)}\text{grad}\,F(\mathbf{x}_{(i)}) - \mathbf{x}^*\right)^T \mathbf{C}\left(\mathbf{x}_{(i)} + \lambda_{(i)}\text{grad}\,F(\mathbf{x}_{(i)}) - \mathbf{x}^*\right) \\ &= \mathbf{c}^T\left(\mathbf{x}_{(i)} + \lambda_{(i)}\text{grad}\,F(\mathbf{x}_{(i)}) - \mathbf{x}^*\right) \\ &\quad + \left(\mathbf{x}_{(i)} + \lambda_{(i)}\text{grad}\,F(\mathbf{x}_{(i)}) - \mathbf{x}^*\right)^T \tilde{\mathbf{C}}\left(\mathbf{x}_{(i)} + \lambda_{(i)}\text{grad}\,F(\mathbf{x}_{(i)}) - \mathbf{x}^*\right) \end{aligned} \tag{3.64}$$

The necessary and sufficient extremum condition for the quadratic function, i.e., $p'\left(\lambda^*_{(i)}\right) \stackrel{!}{=} 0$, yields

$$\begin{aligned} \lambda^*_{(i)} &= -\frac{\left[\text{grad}\,F(\mathbf{x}_{(i)})\right]^T \text{grad}\,F(\mathbf{x}_{(i)})}{2\left[\text{grad}\,F(\mathbf{x}_{(i)})\right]^T \mathbf{C}\text{grad}\,F(\mathbf{x}_{(i)})} \\ &= -\frac{\left[\text{grad}\,F(\mathbf{x}_{(i)})\right]^T \text{grad}\,F(\mathbf{x}_{(i)})}{\left[\text{grad}\,F(\mathbf{x}_{(i)})\right]^T \mathbf{H}(F(\mathbf{x}))\text{grad}\,F(\mathbf{x}_{(i)})} \end{aligned} \tag{3.65}$$

A description of two quasi-Newton methods follows.

Davidon–Fletcher–Powell Method. This method combines the quasi-Newton principle (3.61) of estimating the Hessian (or, more precisely, its inverse) by a sequence $\left\{\mathbf{P}_{(i)}\right\}_{i=1}^{N} = \left\{-\mathbf{S}_{(i)}^{-1}\right\}_{i=1}^{N}$ and the long-step steeping descent method. The algorithm for searching a minimum is presented.

Algorithm 3.11 (Davidon–Fletcher–Powell method)

1. Let objective function $f(\mathbf{x})$ (the gradient of which exists) and the initial extremum estimate $\mathbf{x}_{(1)} \in \mathbb{R}^n$ be given. Select the desired accuracy $\varepsilon > 0$ and gradient norm threshold $\varepsilon \gg e > 0$. Set $i = 1$ and $\mathbf{P}_{(i)} = \mathbf{I}$.
2. Calculate $\text{grad} f(\mathbf{x})$.
3. Compute $\text{grad} f(\mathbf{x}_{(i)})$. If $\left\|\text{grad} f(\mathbf{x}_{(i)})\right\| < e$, set $i = i - 1$ and go to step 7; else, compute

$$\mathbf{d}_{(i)} = \mathbf{P}_{(i)}\text{grad} f(\mathbf{x}_{(i)}) \tag{3.66}$$

   and go to step 4.
4. Construct function

$$p(\lambda_{(i)}) := f(\mathbf{x}_{(i)} + \lambda_{(i)}\mathbf{d}_{(i)}) \tag{3.67}$$

   and solve one-dimensional optimization subproblem (3.60) with the constraint condition $\lambda^*_{(i)} < 0$ (for a minimum search).

5. Compute a new solution estimate

$$\mathbf{x}_{(i+1)} = \mathbf{x}_{(i)} + \lambda^*_{(i)}\mathbf{d}_{(i)} \tag{3.68}$$

6. If (3.57) does not hold, update matrix $\mathbf{P}$ using the following formulae

$$\mathbf{P}_{(i+1)} = \mathbf{P}_{(i)} + \frac{\mathbf{v}_{(i)}\mathbf{v}_{(i)}^T}{\mathbf{v}_{(i)}^T\mathbf{v}_{(i)}} - \frac{\mathbf{P}_{(i)}\mathbf{w}_{(i)}\mathbf{w}_{(i)}^T\mathbf{P}_{(i)}}{\left(\mathbf{P}_{(i)}\mathbf{w}_{(i)}\right)^T\mathbf{w}_{(i)}}$$
$$\mathbf{v}_{(i)} = \mathbf{x}_{(i+1)} - \mathbf{x}_{(i)},\ \mathbf{w}_{(i)} = \operatorname{grad} f\left(\mathbf{x}_{(i+1)}\right) - \operatorname{grad} f\left(\mathbf{x}_{(i)}\right) \tag{3.69}$$

set $i = i + 1$, and go to step 3. Otherwise, go to step 7.

7. Stop and return $\mathbf{x}^* = \mathbf{x}_{(i+1)}$.

Compared to the Newton method, the advantage of this method is that it is not necessary to calculate the Hessian at each step. The method converges to a global extremum for convex functions, and finds an exact analytical solution for quadratic functions.

A Matlab code for a two-variable case is sketched, followed by an illustrative example.

```
% INPUTS
syms f x1 x2 lam
f = input('Enter the objective function: ');
x_init = input('Enter the initial point [x1 x2]: ');
e = input('Enter the desired accuracy: ');

% INIT
n = 2;
grad_t = gradient(f);
x_prev = x_init';
x = x_init';
dx = inf;
P = eye(n);
grad_fx = subs(grad_f,[x1;x2],x);


% ITERATIONS
while dx >= e
   if eval(norm(grad_fx)<eps)
      break
   end
   p = subs(f,[x1;x2],x+lam*P*grad_fx);
   % ONE-DIMENSIONAL OPTIMIZATION
   dp = diff(p,lam);
   lam0 = solve(dp==0);
   n_lam0 = length(lam0);
   lam_min=inf;
   for j = 1:1:n_lam0
      l = 2;
```

```
        while eval(subs(diff(dp,lam),lam,lam0(j))==0)
           dp = diff(dp,lam);
           l = l+1;
        end
       if (rem(l,2)==0) && eval(subs(diff(dp,lam),lam,lam0(j))>0) &&
   (lam0(j)<0)
           if lam0(j) < lam_min
              lam_min = lam0(j); % A BETTER SOLUTION FOUND
           end
        end
     end
     x = x+lam_min*P*grad_fx;

     % UPDATE MATRIX P
     grad_prev = grad_fx;
     grad_fx = subs(grad_f,[x1;x2],x);
     v = x-x_prev;
     w = grad_fx-grad_prev;
     P = P+v*v.'/(v.'*v)-P*w*w.'*P/((P*w)'*w);

     dx = norm(v);
     x_prev = x;
  end

  % OUTPUT
  x
```

Example 3.11 *Task*: Find a minimum of $f(\mathbf{x}) = f(x_1, x_2) = 3x_1^2 - 2x_1x_2 + x_2^2 + 4x_1 + 3x_2 + 1$ on $\mathbb{R}^2$ using the Davidon–Fletcher–Powell method starting from the initial estimate $\mathbf{x}_{(1)} = [0, 0]^T$. Let the given accuracies be $\varepsilon = 0.05$, $e = 5 \times 10^{-3}$.

Solution: As $\mathbf{P}_{(i)} = \mathbf{I}$, the initial solution direction equals the gradient of $f(\mathbf{x})$ (that already has been calculated in Example 3.9). That is,

$$\mathbf{d}_{(1)} = \mathbf{I}\mathrm{grad} f\left(\mathbf{x}_{(1)}\right) = \mathrm{grad} f\left(\mathbf{x}_{(1)}\right) = [4, 3]^T$$

see (3.66). Then, the long-step steepest descent method in direction $\mathbf{d}_{(1)}$ is applied, see (3.67). Since $\mathbf{d}_{(1)} = \mathrm{grad} f\left(\mathbf{x}_{(1)}\right)$, the new solution estimate for $i = 1$ is identical to that obtained in Example 3.10b, i.e.,

$$\begin{aligned}\mathbf{x}_{(2)} &= \mathbf{x}_{(1)} + \lambda_{(i)}^* \mathbf{d}_{(1)} = [0, 0]^T - \frac{25}{66}[4, 3]^T = [-50/33, -25/22]^T \\ &= [-1.5251, -1.1364]^T\end{aligned}$$

Now, $\mathrm{grad} f\left(\mathbf{x}_{(2)}\right) = [-2.8182, 3.7576]^T$ is computed (which norm is apparently higher than e), and the value of $\mathbf{P}_{(i)} = \mathbf{I}$ is updated as per (3.69)

$$\mathbf{v}_{(1)} = \mathbf{x}_{(2)} - \mathbf{x}_{(1)} = [-1.5152, -1.1364]^T$$
$$\mathbf{w}_{(1)} = \operatorname{grad} f\left(\mathbf{x}_{(2)}\right) - \operatorname{grad} f\left(\mathbf{x}_{(1)}\right) = [-6.8182, 0.7576]^T$$
$$\mathbf{P}_{(2)} = \mathbf{P}_{(1)} + \frac{\mathbf{v}_{(1)}\mathbf{v}_{(1)}^T}{\mathbf{v}_{(1)}^T\mathbf{v}_{(1)}} - \frac{\mathbf{P}_{(1)}\mathbf{w}_{(1)}\mathbf{w}_{(1)}^T\mathbf{P}_{(1)}}{\left(\mathbf{P}_{(1)}\mathbf{w}_{(1)}\right)^T\mathbf{w}_{(1)}} = \begin{bmatrix} 0.6522 & 0.5898 \\ 0.5898 & 1.3478 \end{bmatrix}$$

Update the iteration step number $i = 2$.

It is worth noting that the optimal $\lambda^*_{(i)}$ can be obtained analytically for this example. Hence, denote

$$\mathbf{P}_{(i)} = \begin{bmatrix} p_{11,(i)} & p_{12,(i)} \\ p_{21,(i)} & p_{22,(i)} \end{bmatrix}$$

Then, the solution of the optimization subproblem (3.60) with (3.67) is given by applying condition $p'\left(\lambda^*_{(i)}\right) \overset{!}{=} 0$ as

$$\lambda^*_{(i)} = \frac{\begin{aligned} &\left(-36p_{11,(i)} + 12p_{12,(i)} + 12p_{12,(i)} - 4p_{22,(i)}\right)x_{1,(i)}^2 \\ &+\left(-4p_{11,(i)} + 4p_{12,(i)} + 4p_{12,(i)} - 4p_{22,(i)}\right)x_{2,(i)}^2 \\ &+\left(-48p_{11,(i)} - 10p_{12,(i)} - 10p_{12,(i)} + 12p_{22,(i)}\right)x_{1,(i)} \\ &+\left(16p_{11,(i)} - 2p_{12,(i)} - 2p_{12,(i)} - 12p_{22,(i)}\right)x_{2,(i)} \\ &+\left(24p_{11,(i)} - 16p_{12,(i)} - 16p_{12,(i)} + 8p_{22,(i)}\right)x_{1,(i)}x_{2,(i)} \\ &-16p_{11,(i)} - 12p_{12,(i)} - 12p_{12,(i)} - 9p_{22,(i)} \end{aligned}}{\begin{aligned} &6\left(p_{11,(i)}\left(6x_{1,(i)} - 2x_{2,(i)} + 4\right) + p_{12,(i)}\left(-2x_{1,(i)} + 2x_{2,(i)} + 3\right)\right)^2 \\ &-4\left(p_{11,(i)}\left(6x_{1,(i)} - 2x_{2,(i)} + 4\right) + p_{12,(i)}\left(-2x_{1,(i)} + 2x_{2,(i)} + 3\right)\right) \\ &\left(p_{21,(i)}\left(6x_{1,(i)} - 2x_{2,(i)} + 4\right) + p_{22,(i)}\left(-2x_{1,(i)} + 2x_{2,(i)} + 3\right)\right) \\ &+2\left(p_{21,(i)}\left(6x_{1,(i)} - 2x_{2,(i)} + 4\right) + p_{22,(i)}\left(-2x_{1,(i)} + 2x_{2,(i)} + 3\right)\right)^2 \end{aligned}}$$

By substituting $\mathbf{P}_{(2)}$, $\mathbf{x}_{(2)}$ into the result above, one gets $\lambda^*_{(2)} = 44/61$. Taking (3.68), the new solution estimate is obtained

$$\begin{aligned} \mathbf{x}_{(3)} &= \mathbf{x}_{(2)} + \lambda^*_{(2)}\operatorname{grad} f\left(\mathbf{x}_{(2)}\right) \\ &= [-1.5251, -1.1364]^T - \frac{44}{61}\begin{bmatrix} 0.6522 & 0.5898 \\ 0.5898 & 1.3478 \end{bmatrix}[-2.8182, 3.7576]^T \\ &= [-1.7501, -3.2501]^T \end{aligned}$$

The corresponding gradient reads $\operatorname{grad} f\left(\mathbf{x}_{(3)}\right) = [-0.005, 0.0001]^T$. As $\left\|\operatorname{grad} f\left(\mathbf{x}_{(3)}\right)\right\|_2 = 1.584 \times 10^{-3} < 5 \times 10^{-3} = e$, the calculation is finished. Besides, the obtained minimum estimate $\mathbf{x}_{(3)}$ is sufficiently close to the actual extremum $\mathbf{x}^* = [-1.75, -3.25]^T$.

Hence (as expected), the method converges to the extremum for a quadratic function in one step after updating **P** since it is based on the Newton method.

Conjugate Gradient Method. Although this method is usually excluded from the family of quasi-Newton methods, it adopts some of their shared principles; therefore, it is placed in this section.

Conjugate directions are defined first.

Definition 3.3 Given a positive definite matrix $\mathbf{M} \in \mathbb{R}^{n \times n}$, it is said that **vectors** $\mathbf{v}_i \in \mathbb{R}^n, i = 1, 2, \ldots, n$, are (mutually) **conjugated with respect to M** if for every $i, j = 1, 2, \ldots, n, i \neq j$, it holds that

$$\mathbf{v}_i^T \mathbf{M} \mathbf{v}_j = 0 \tag{3.70}$$

Note that the conjugate directions generalize the notion of the *orthogonality*.

Consider quadratic (approximating) function $F(\mathbf{x})$ as in (3.62). Then, the following statement holds.

Proposition 3.3 *Let* $\mathbf{v}_i \in \mathbb{R}^n, i = 1, 2, \ldots, n$ *be conjugated with respect to* **C**. *Then, the following sequence of points*

$$\begin{aligned} \mathbf{x}_{(i+1)} &= \mathbf{x}_{(i)} + \lambda_{(i)} \mathbf{v}_i \\ \lambda_{(i)} &= -\frac{\left[\operatorname{grad} F\left(\mathbf{x}_{(i)}\right)\right]^T \mathbf{v}_i}{2\mathbf{v}_i^T \mathbf{C} \mathbf{v}_i} \end{aligned} \tag{3.71}$$

conjugates to a minimum of $F(\mathbf{x})$.

Readers can compare (3.71) with (3.63) and (3.65). I.e., (3.71) becomes by substituting $\operatorname{grad} F\left(\mathbf{x}_{(i)}\right) = \mathbf{v}_i$ and (3.63) into (3.65).

Two algorithms of the conjugate gradient method follow. First, a particular case that converges for quadratic functions is given. Second, a general nonlinear-function algorithm for a minimum search is presented.

Algorithm 3.12a (Conjugate gradient method—the quadratic-function version)

1. Let objective function $f(\mathbf{x})$ (the gradient and Hessian of which exist) and the initial extremum estimate $\mathbf{x}_{(1)} \in \mathbb{R}^n$ be given. Select the desired accuracy $\varepsilon > 0$ and gradient norm threshold $\varepsilon \gg e > 0$. Set $i = 1$.
2. Calculate $\operatorname{grad} f(\mathbf{x})$ and the Hessian $\mathbf{H}(f(\mathbf{x}))$.
3. Compute $\operatorname{grad} f\left(\mathbf{x}_{(i)}\right)$ and $\mathbf{H}\left(f\left(\mathbf{x}_{(i)}\right)\right)$. If $\left\|\operatorname{grad} f\left(\mathbf{x}_{(i)}\right)\right\| < e$, set $i = i - 1$ and go to step 8; else, go to step 4
4. If $i = 1$, set $\mathbf{d}_{(i)} = \operatorname{grad} f\left(\mathbf{x}_{(i)}\right)$; else, set

$$\mathbf{d}_{(i)} = \operatorname{grad} f\left(\mathbf{x}_{(i)}\right) - \frac{\left[\operatorname{grad} f\left(\mathbf{x}_{(i)}\right)\right]^T \mathbf{H}\left(f\left(\mathbf{x}_{(i)}\right)\right) \mathbf{d}_{(i-1)}}{\mathbf{d}_{(i-1)}^T \mathbf{H}\left(f\left(\mathbf{x}_{(i-1)}\right)\right) \mathbf{d}_{(i-1)}} \mathbf{d}_{(i-1)} \tag{3.72}$$

5. Compute

$$\lambda_{(i)}^{*} = -\frac{\left[\operatorname{grad} f\left(\mathbf{x}_{(i)}\right)\right]^{T} \mathbf{d}_{(i)}}{\mathbf{d}_{(i)}^{T}\, \mathbf{H}\left(f\left(\mathbf{x}_{(i)}\right)\right) \mathbf{d}_{(i)}} \tag{3.73}$$

6. Compute a new solution estimate $\mathbf{x}_{(i+1)}$ via (3.68).
7. If (3.57) does not hold, set $i = i + 1$ and go to step 3. Otherwise, go to step 8.
8. Stop and return $\mathbf{x}^{*} = \mathbf{x}_{(i+1)}$.

Algorithm 3.12b (Conjugate gradient method—the general-function version)

1. Let objective function $f(\mathbf{x})$ (the gradient of which exists) and the initial extremum estimate $\mathbf{x}_{(1)} \in \mathbb{R}^n$ be given. Select the desired accuracy $\varepsilon > 0$ and gradient norm threshold $\varepsilon \gg e > 0$. Set $i = 1$.
2. Calculate $\operatorname{grad} f(\mathbf{x})$.
3. Compute $\operatorname{grad} f\left(\mathbf{x}_{(i)}\right)$. If $\left\|\operatorname{grad} f\left(\mathbf{x}_{(i)}\right)\right\| < e$, set $i = i - 1$ and go to step 8; else, go to step 4.
4. If $i = 1$, set $\mathbf{d}_{(i)} = \operatorname{grad} f\left(\mathbf{x}_{(i)}\right)$; else, set

$$\mathbf{d}_{(i)} = \operatorname{grad} f\left(\mathbf{x}_{(i)}\right) + \frac{\left[\operatorname{grad} f\left(\mathbf{x}_{(i)}\right)\right]^{T} \operatorname{grad} f\left(\mathbf{x}_{(i)}\right)}{\left[\operatorname{grad} f\left(\mathbf{x}_{(i-1)}\right)\right]^{T} \operatorname{grad} f\left(\mathbf{x}_{(i-1)}\right)} \mathbf{d}_{(i-1)} \tag{3.74}$$

5. Construct function $p\left(\lambda_{(i)}\right)$ as in (3.67) and solve one-dimensional optimization subproblem (3.60) with the constraint condition $\lambda_{(i)}^{*} < 0$ (for a minimum search).
6. Compute a new solution estimate $\mathbf{x}_{(i+1)}$ via (3.68).
7. If (3.57) does not hold, set $i = i + 1$ and go to step 3. Otherwise, go to step 8.
8. Stop and return $\mathbf{x}^{*} = \mathbf{x}_{(i+1)}$.

Possible Matlab codes (for a two-variable case) of Algorithms 3.12a and 3.12b, respectively, follow.

```
% INPUTS
syms f x1 x2 lam
f = input('Enter the objective function: ');
x_init = input('Enter the initial point [x1 x2]: ');
e = input('Enter the desired accuracy: ');

% INIT
n = 2;
i = 1;
grad_f = gradient(f);
H_f = hessian(f);
x = x_init';
dx = inf;


% ITERATIONS
```

```
while dx >= e
   grad_fx = subs(grad_f,[x1;x2],x);
   H_fx = subs(H_f,[x1;x2],x);
   if eval(norm(grad_fx)<eps)
      break
   end
   if i == 1
      d = grad_fx;
   else
         d = grad_fx-grad_fx.'*H_fx*d_prev/(d_prev.'*H_prev*d_
prev)*d_prev;
   end
   lam_min = -grad_fx.'*d/(d.'*H_fx*d);
   x = x+lam_min*d;

   dx = norm(lam_min*d);
   d_prev = d;
   H_prev = H_fx;
   i=i+1;
end

% OUTPUT
x

% INPUTS
syms f x1 x2 lam
f = input('Enter the objective function: ');
x_init = input('Enter the initial point [x1 x2]: ');
e = input('Enter the desired accuracy: ');

% INIT
n = 2;
i = 1;
grad_f = gradient(f);
x = x_init';
dx = inf;


% ITERATIONS
while dx >= e
   grad_fx = subs(grad_f,[x1;x2],x);
   if eval(norm(grad_fx)<eps)
      break
   end
   if i == 1
      d = grad_fx;
   else
        d = grad_fx-grad_fx.'*grad_fx/(grad_prev.'*grad_prev)*d_
prev;
   end
   p = subs(f,[x1;x2],x+lam*d);
   % ONE-DIMENSIONAL OPTIMIZATION
```

```
    dp = diff(p,lam);
    lam0 = solve(dp==0);
    n_lam0 = length(lam0);
    lam_min=inf;
    for j = 1:1:n_lam0
      l = 2;
      while eval(subs(diff(dp,lam),lam,lam0(j))==0)
        dp = diff(dp,lam);
        l = l+1;
      end
     if (rem(l,2)==0) && eval(subs(diff(dp,lam),lam,lam0(j))>0) &&
(lam0(j)<0)
        if lam0(j) < lam_min
          lam_min = lam0(j); % A BETTER SOLUTION FOUND
        end
      end
    end
    dx = lam_min*d;
    x = x+dx;

    dx = norm(dx);
    grad_prev = grad_fx;
    d_prev = d;
    i = i+1;
end

% OUTPUT
x
```

An example illustrating how the two algorithms work is given to readers.

Example 3.12 *Task*: Find a minimum of $f(\mathbf{x}) = f(x_1, x_2) = 3x_1^2 - 2x_1x_2 + x_2^2 + 4x_1 + 3x_2 + 1$ on $\mathbb{R}^2$ using the conjugate gradient method starting from the initial estimate $\mathbf{x}_{(1)} = [0, 0]^T$. Let the given accuracies be $\varepsilon = 0.05$, $e = 5 \times 10^{-3}$. Compare the quadratic-function and general-function methods (variants).

Solution: Let the quadratic-function method be considered first. Besides the gradient, it is necessary to calculate also the Hessian that is known from Example 3.9. Note that the gradient can be found therein as well.

The optimal initial step length reads

$$\lambda^*_{(1)} = -\frac{\left[\operatorname{grad} f\left(\mathbf{x}_{(1)}\right)\right]^T \mathbf{d}_{(1)}}{\mathbf{d}^T_{(1)}\,\mathbf{H}\big(f\left(\mathbf{x}_{(1)}\right)\big)\,\mathbf{d}_{(1)}}$$
$$= -\frac{\left[\operatorname{grad} f\left(\mathbf{x}_{(1)}\right)\right]^T \operatorname{grad} f\left(\mathbf{x}_{(1)}\right)}{\left[\operatorname{grad} f\left(\mathbf{x}_{(1)}\right)\right]^T \mathbf{H}\big(f\left(\mathbf{x}_{(1)}\right)\big) \operatorname{grad} f\left(\mathbf{x}_{(1)}\right)}$$
$$= -\frac{[4, 3][4, 3]^T}{[4, 3]\begin{bmatrix} 6 & -2 \\ -2 & 2 \end{bmatrix}[4, 3]^T} = -0.3788$$

Apparently, the value of $\lambda^*_{(1)}$ is identical to that in Example 3.10 for the long-step steeping descent method; therefore, $\mathbf{x}_{(2)} = [-1.5152, -1.1364]^T$, $\operatorname{grad} f(\mathbf{x}_{(2)}) = [-2.8182, 3.7576]^T$. Then, a new conjugate direction reads

$$\begin{aligned}
\mathbf{d}_{(2)} &= \operatorname{grad} f(\mathbf{x}_{(2)}) - \frac{[\operatorname{grad} f(\mathbf{x}_{(2)})]^T \mathbf{H}(f(\mathbf{x}_{(2)}))\mathbf{d}_{(1)}}{\mathbf{d}_{(1)}^T \mathbf{H}(f(\mathbf{x}_{(1)}))\, \mathbf{d}_{(1)}} \mathbf{d}_{(1)} \\
&= [-2.8182, 3.7576]^T - \frac{[-2.8182, 3.7576]\begin{bmatrix} 6 & -2 \\ -2 & 2 \end{bmatrix}[4, 3]^T}{[4, 3]\begin{bmatrix} 6 & -2 \\ -2 & 2 \end{bmatrix}[4, 3]^T}[4, 3]^T \\
&= [0.7117, 6.4050]^T
\end{aligned}$$

The optimal step length $\lambda^*_{(2)}$ at $\mathbf{x}_{(2)}$ is

$$\begin{aligned}
\lambda^*_{(2)} &= -\frac{[\operatorname{grad} f(\mathbf{x}_{(2)})]^T \mathbf{d}_{(2)}}{\mathbf{d}_{(2)}^T \mathbf{H}(f(\mathbf{x}_{(2)}))\, \mathbf{d}_{(2)}} \\
&= -\frac{[-2.8182, 3.7576][0.7117, 6.4050]^T}{[0.7117, 6.4050]\begin{bmatrix} 6 & -2 \\ -2 & 2 \end{bmatrix}[0.7117, 6.4050]^T} = -0.33
\end{aligned}$$

The corresponding updated solution estimate reads

$$\begin{aligned}
\mathbf{x}_{(3)} &= \mathbf{x}_{(2)} + \lambda^*_{(2)}\mathbf{d}_{(2)} = [-1.5251, -1.1364]^T - 0.33[0.7117, 6.4050]^T \\
&= [-1.7501, -3.2501]^T
\end{aligned}$$

It is clear that this point is very close to that obtained using the Davidon–Fletcher–Powell method in Example 3.11 at the same iteration step.

Now, let the general (nonlinear) method be applied. For $i = 1$, $\mathbf{d}_{(1)} = \operatorname{grad} f(\mathbf{x}_{(1)}) = [4, 3]^T$; therefore, $\lambda^*_{(1)} = -25/66 = -0.3788$ again. Generally, it is possible to calculate the optimal $\lambda^*_{(i)}$ as

$$\begin{aligned}
\lambda^*_{(i)} &= \arg\min_{\lambda \in \mathbb{R}} p(\lambda_{(i)}) \\
p(\lambda_{(i)}) &= f(\mathbf{x}_{(i)} + \lambda_{(i)}\mathbf{d}_{(i)}) = f\left([x_{1,(i)}, x_{2,(i)}]^T + \lambda_{(i)}[d_{1,(i)}, d_{2,(i)}]^T\right) \\
&= 3(x_{1,(i)} + \lambda_{(i)}d_{1,(i)})^2 - 2(x_{1,(i)} + \lambda_{(i)}d_{1,(i)})(x_{2,(i)} + \lambda_{(i)}d_{2,(i)}) \\
&\quad + (x_{2,(i)} + \lambda_{(i)}d_{2,(i)})^2 + 4(x_{1,(i)} + \lambda_{(i)}d_{1,(i)}) + 3(x_{2,(i)} + \lambda_{(i)}d_{2,(i)}) + 1 \\
p'(\lambda_{(i)}) &\stackrel{!}{=} 0 \Rightarrow \lambda^*_{(i)} = 0.5\frac{d_{1,(i)}(-6x_{1,(i)} + 2x_{2,(i)} - 4) + d_{2,(i)}(2x_{1,(i)} - 2x_{2,(i)} - 3)}{3d^2_{1,(i)} + d^2_{2,(i)} - 2d^2_{1,(i)}d^2_{2,(i)}}
\end{aligned}$$

Hence, $\mathbf{x}_{(2)} = [-1.5152, -1.1364]^T$, $\operatorname{grad} f(\mathbf{x}_{(2)}) = [-2.8182, 3.7576]^T$. The new conjugated direction is computed from

$$\begin{aligned}
\mathbf{d}_{(2)} &= \operatorname{grad} f(\mathbf{x}_{(2)}) + \frac{\left[\operatorname{grad}\left(f(\mathbf{x}_{(2)})\right)\right]^T \operatorname{grad}\left(f(\mathbf{x}_{(2)})\right)}{\left[\operatorname{grad}\left(f(\mathbf{x}_{(1)})\right)\right]^T \operatorname{grad}\left(f(\mathbf{x}_{(1)})\right)} \mathbf{d}_{(1)} \\
&= [-2.8182, 3.7576]^T + \frac{[-2.8182, 3.7576][-2.8182, 3.7576]^T}{[4, 3][4, 3]^T} [4, 3]^T \\
&= [0.7117, 6.4050]^T
\end{aligned}$$

The obtained result is identical to that from the quadratic method, which is plain to see. This is because the objective function is quadratic.

Intuitively, it is clear why the conjugate gradient method (even the long-step steepest descent one) cannot converge to a solution in a single iterative step. Namely, all elements of vector $\mathbf{d}$ (or $\operatorname{grad} f(\mathbf{x})$) are multiplied only by λ^*, which is a scalar. In contrast, for example, in the Newton method or the Davidon–Fletcher–Powell method, a matrix is placed instead of a scalar. That is, each coordinate of the current solution estimate is modified by a linear combination of gradient elements (which apparently entails greater variability).

3.2.2.4 Parallel Tangent Method

The last method from the family of multi-dimensional gradient-based presented in this book is the parallel tangent method. Its principle resides in a multiple use of the (long-step) steepest descent method and the "averaging" of the obtained subresults within a particular iteration step.

In principle, there are two nested loops here. The outer one goes through the iteration. The inner loop performs long-step movements in the gradient direction, the number of which equals the variable space dimension (n). This principle is expressed graphically in Fig. 3.17.

A sketch of the parallel tangent method follows.

Algorithm 3.13 (Parallel tangent method)

1. Let objective function $f(\mathbf{x})$ (the gradient of which exists) and the initial extremum estimate $\mathbf{x}_{(1)} \in \mathbb{R}^n$ be given. Select the desired accuracy $\varepsilon > 0$. Set $i = j = 1$ and $\mathbf{x}_{(i),(j)} = \mathbf{x}_{(1)}$.
2. Calculate $\operatorname{grad} f(\mathbf{x})$.
3. Compute $\operatorname{grad} f(\mathbf{x}_{(i),(j)})$ and solve the optimization subproblem

$$\begin{aligned}
\lambda^*_{(i),(j)} &= \arg\operatorname{extr} p(\lambda_{(i),(j)}) \\
p(\lambda_{(i),(j)}) &:= f\left(\mathbf{x}_{(i),(j)} + \lambda_{(i),(j)} \operatorname{grad} f(\mathbf{x}_{(i),(j)})\right)
\end{aligned} \tag{3.75}$$

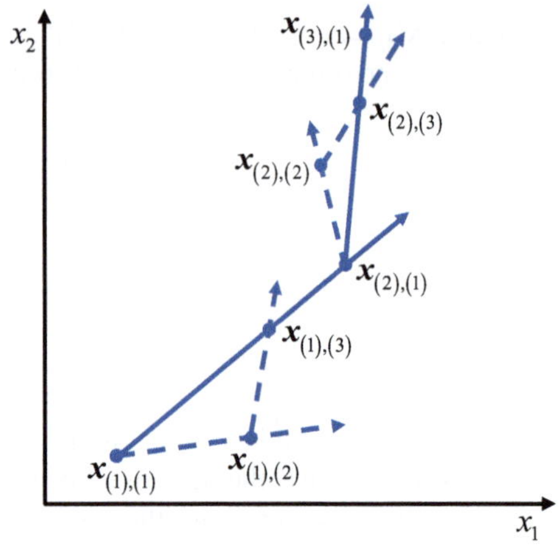

Fig. 3.17 A principle of the parallel tangent method

and update the extremum estimate within the inner loop

$$\mathbf{x}_{(i),(j+1)} = \mathbf{x}_{(i),(j)} + \lambda^*_{(i),(j)} \operatorname{grad} f\left(\mathbf{x}_{(i),(j)}\right) \tag{3.76}$$

4. If $j \leq n$, set $j = j + 1$, and go to step 3. Otherwise, set $j = 1$, and go to step 5.
5. Solve the optimization subproblem

$$\begin{aligned} \lambda^*_{(i)} &= \arg\operatorname{extr} p\left(\lambda_{(i)}\right) \\ p\left(\lambda_{(i)}\right) &:= f\left(\mathbf{x}_{(i),(j)} + \lambda_{(i)}\left(\mathbf{x}_{(i),(n+1)} - \mathbf{x}_{(i),(j)}\right)\right) \end{aligned} \tag{3.77}$$

and update the extremum estimate within the inner loop

$$\begin{aligned} \mathbf{x}_{(i+1),(j)} &= \mathbf{x}_{(i),(j)} + \lambda^*_{(i)}\left(\mathbf{x}_{(i),(n+1)} - \mathbf{x}_{(i),(j)}\right) \\ &= \lambda^*_{(i)}\mathbf{x}_{(i),(n+1)} + \left(1 - \lambda^*_{(i)}\right)\mathbf{x}_{(i),(j)} \end{aligned} \tag{3.78}$$

6. If $\left\|\mathbf{x}_{(i+1),(j)} - \mathbf{x}_{(i),(j)}\right\| \geq \varepsilon$, set $i = i + 1$ and go to step 3; else, go to step 7.
7. Stop and return $\mathbf{x}^* = \mathbf{x}_{(i+1),(j)}$.

In step 3 of Algorithm 3.13, it is implicitly supposed that the algorithm is terminated if the gradient norm is "low"; see step 3 of Algorithms 3.10a, 3.10b, 3.11, 3.12a, and 3.12b. This condition is omitted due to clarity.

Note that when minimizing $f(\mathbf{x})$, it should hold that $\lambda^*_{(i),(j)} < 0$ in (3.75), as it indicates the function value decreases in the gradient direction. Contrariwise, it is expected that $\lambda^*_{(i)} > 0$ in (3.77) since the given direction $\mathbf{x}_{(i),(n+1)} - \mathbf{x}_{(i),(j)}$ is towards a minimum.

A Matlab code for a 2-variable case when searching for a minimum can be performed as follows.

```
% INPUTS
syms f x1 x2 lam
f = input('Enter the objective function: ');
x_init = input('Enter the initial point [x1 x2]: ');
e = input('Enter the desired accuracy: ');

% INIT
n = 2;
j = 1;
grad_f = gradient(f);
x = x_init';
x_start = x;
dx = inf;

% ITERATIONS
while dx >= e
   for j = 1:1:n
      grad_fx = subs(grad_f,[x1;x2],x);
      if eval(norm(grad_fx)<eps)
         x
         return
      end
      p = subs(f,[x1;x2],x+lam*grad_fx);
      lam_min = lamMin(p,lam);
      x = x+lam_min*grad_fx;
   end
   p = subs(f,[x1;x2],x_start+lam*(x-x_start));
   lam_max = lamMax(p,lam);
   x = x_start+lam_max*(x-x_start);

   x = norm(x-x_start);
   x_start = x;
end

% OUTPUT
x
```

where the following functions are used:

```
function lam_min = lamMin(p,lam)
   % ONE-DIMENSIONAL OPTIMIZATION
   dp = diff(p,lam);
   lam0 = solve(dp==0,lam);
   n_lam0 = length(lam0);
   lam_min=inf;
   for k = 1:1:n_lam0
      l = 2;
      while eval(subs(diff(dp,lam),lam,lam0(k))==0)
```

```
            dp = diff(dp,lam);
            l = l+1;
         end
       if (rem(l,2)==0) && eval(subs(diff(dp,lam),lam,lam0(k))>0) &&
(lam0(k)<0)
            if lam0(k) < lam_min
               lam_min = lam0(k); % A BETTER SOLUTION FOUND
            end
         end
      end
   end

   function lam_max = lamMax(p,lam)
      % ONE-DIMENSIONAL OPTIMIZATION
      dp = diff(p,lam);
      lam0 = solve(dp==0,lam);
      n_lam0 = length(lam0);
      lam_max=-inf;
      for k = 1:1:n_lam0
         l = 2;
         while eval(subs(diff(dp,lam),lam,lam0(k))==0)
            dp = diff(dp,lam);
            l = l+1;
         end
       if (rem(l,2)==0) && eval(subs(diff(dp,lam),lam,lam0(k))>0) &&
(lam0(k)>0)
            if lam0(k) > lam_max
               lam_max = lam0(k); % A BETTER SOLUTION FOUND
            end
         end
      end
   end
```

A numerical example follows.

Example 3.13 *Task*: Find a minimum of $f(\mathbf{x}) = f(x_1, x_2) = 3x_1^2 - 2x_1x_2 + x_2^2 + 4x_1 + 3x_2 + 1$ on $\mathbb{R}^2$ using the conjugate gradient method starting from the initial estimate $\mathbf{x}_{(1)} = [0, 0]^T$. Let the given accuracies be $\varepsilon = 0.05$, $e = 5 \times 10^{-3}$ again.

Solution: Set $i = j = 1$ and $\mathbf{x}_{(i),(j)} = \mathbf{x}_{(1),(1)} = [0, 0]^T$. Function grad$f(\mathbf{x})$ and its value at $\mathbf{x}_{(1),(1)}$ are already known from Example 3.9. Hence, $\mathbf{x}_{(1),(2)}$ is computed exactly as for the initial iteration step of the long-step steepest descent algorithm, i.e.

$$\mathbf{x}_{(1),(2)} = \mathbf{x}_{(1),(1)} + \lambda^*_{(1),(1)} \mathrm{grad} f\left(\mathbf{x}_{(1),(1)}\right) = [-1.5152, -1.1364]$$

with $\lambda^*_{(1),(1)} = 25/66$. The inner loop then continues by computing another long-step movement in the direction of the current gradient grad$f\left(\mathbf{x}_{(1),(2)}\right) = [-2.8182, 3.7576]^T$. The optimal multiplicative constant $\lambda^*_{(1),(2)}$ coincides with the long-step method again, hence

$$\begin{aligned}\mathbf{x}_{(1),(3)} &= \mathbf{x}_{(1),(2)} + \lambda^*_{(1),(2)} \operatorname{grad} f\left(\mathbf{x}_{(1),(2)}\right) \\ &= [-1.5152, -1.1364]^T - 0.1866[-2.8182, 3.7576]^T \\ &= [-50/33, -25/22]^T = [-0.9894, -1.8374]^T\end{aligned}$$

Now, set $j = 1$, and both the gradient-direction steps are "averaged" using (3.77), i.e., an optimal movement in the $\mathbf{x}_{(1),(3)} - \mathbf{x}_{(1),(1)}$ direction is made. The optimal step length is

$$\lambda^*_{(i)} = \arg\min f\left(\mathbf{x}_{(1),(1)} + \lambda^*_{(i)}\left(\mathbf{x}_{(1),(3)} - \mathbf{x}_{(1),(1)}\right)\right)$$

As in Example 3.12, a general optimal solution can be found. Namely, denote $\mathbf{x}_{(i),(1)} = \left[x_{1,(i)}, x_{2,(i)}\right]^T$, $\mathbf{x}_{(i),(3)} - \mathbf{x}_{(i),(1)} = \left[d_{1,(i)}, d_{2,(i)}\right]^T$, and use the result derived for $\lambda^*_{(i)}$ in Example 3.12.

$$\lambda^*_{(1)} = 0.5\frac{(-0.9894)(-4) + (-1.8374)(-3)}{3(-0.9894)^2 + (-1.8374)^2 - 2(-0.9894)(-1.8374)} = 1.7732$$

Then, (3.78) yields

$$\begin{aligned}\mathbf{x}_{(2),(1)} &= \mathbf{x}_{(1),(1)} + \lambda^*_{(1)}\left(\mathbf{x}_{(1),(3)} - \mathbf{x}_{(1),(1)}\right) \\ &= [0, 0]^T + 1.7732[-0.9894, -1.8374]^T = [-1.7438, -3.2581]^T\end{aligned}$$

Since the obtained result is sufficiently close to the actual extremum (the 2-norm of the error is 1.02×10^{-2}), the computation is stopped. In fact, it must hold that $\left\|\mathbf{x}_{(2),(2)} - \mathbf{x}_{(2),(1)}\right\| < 1.02 \times 10^{-2} < \varepsilon = 5 \times 10^{-2}$.

3.3 Multi-dimensional Constrained Problem

The objective function without limiting conditions has been focused in this chapter so far; i.e., the solution was sought in the n-dimensional space of real numbers. However, numerous real-life problems lead to models with constraints, both in the form of linear and non-linear algebraic equations or even inequalities.

In this section, one iterative algorithm for finding the extremum of a (generally non-linear) multi-variable function with linear inequality constraints is presented, and the idea of using penalty or barrier functions to incorporate the constraints directly into the objective function will also be indicated.

Recall that the analytical solution of the problem with auxiliary conditions in the form of equality was provided by the Lagrange multipliers theorem (see Theorem 2. 3), while necessary analytic conditions for constraints in the form of inequalities were given by the Karush–Kuhn–Tucker theorem (see Theorem 2.4).

3.3.1 Gradient Projection Method

A problem that can be solved by the gradient projection method can be expressed in the form

$$\begin{aligned} &\max f(\mathbf{x}) \in \mathbb{R}, \mathbf{x} \in \mathbb{R}^n \\ &\quad \mathbf{Ax} \leq \mathbf{b},\ \mathbf{A} \in \mathbb{R}^{m\times n},\ \mathbf{b} \in \mathbb{R}^m \\ &\quad \mathbf{x} \geq \mathbf{0} \end{aligned} \tag{3.79}$$

Task (3.79) often appears in the realm of economics, where a widely pursued aim is to optimize the profit function $f(\mathbf{x})$, which accounts for the production volume of each unique product while factoring in non-negative variables $\mathbf{x}$, such as raw materials, warehouse capacity, and workforce size. A visual depiction of the objective function projected onto a two-variable plane, accompanied by a set of potential constraints, is illustrated in Fig. 3.18.

A set of constraints in (3.79) can also be expressed as

$$\begin{aligned} &\mathbf{Ax} \leq \mathbf{b} \\ &-\mathbf{I}_n\mathbf{x} \leq \mathbf{0} \end{aligned}$$

$$\mathbf{A} = \begin{bmatrix} a_{11} & a_{12} & \cdots & a_{1n} \\ a_{21} & a_{22} & \cdots & a_{2n} \\ \vdots & \vdots & \ddots & \vdots \\ a_{m1} & a_{m2} & \cdots & a_{mn} \end{bmatrix} = \begin{bmatrix} \mathbf{a}_1^T \\ \mathbf{a}_2^T \\ \vdots \\ \mathbf{a}_m^T \end{bmatrix}, \mathbf{b} = \begin{bmatrix} b_1 \\ b_2 \\ \vdots \\ b_m \end{bmatrix}, \mathbf{I}_n = \begin{bmatrix} \mathbf{e}_1^T \\ \mathbf{e}_2^T \\ \vdots \\ \mathbf{e}_n^T \end{bmatrix} \tag{3.80}$$

where $\mathbf{I}_n \in \mathbb{R}^{n\times n}$ is the identity matrix (i.e., $\mathbf{e}_i \in \mathbb{R}^n$ are Euclidean vectors).

The algorithm can mathematically be expressed as follows.

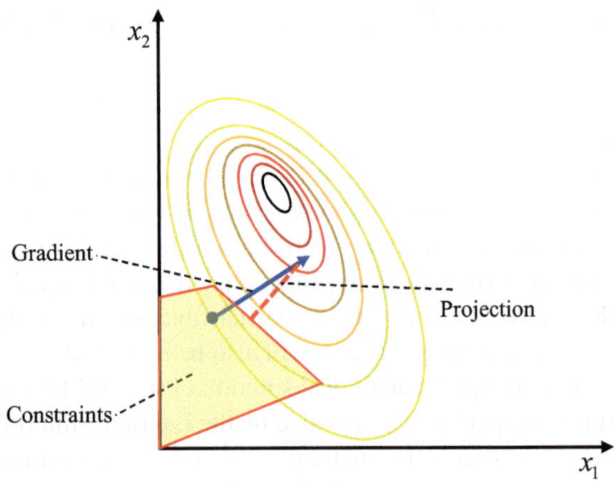

Fig. 3.18 A principle of the gradient projection method

Algorithm 3.14 (Gradient projection method)

1. Let objective function $f(\mathbf{x})$ (the gradient of which exists), a set of linear inequality constraints as in (3.80), and the initial extremum estimate $\mathbf{x}_{(1)} \in \mathbb{R}^n$ be given. The value of $\mathbf{x}_{(1)}$ must satisfy at least one equality in (3.80); i.e., the point must lie at the boundary of the constraint set. Select the desired accuracy $\varepsilon > 0$. Set $i = 1$ and $\mathbf{x}_{(i)} = \mathbf{x}_{(1)}$.
2. Define sets of indexes

$$\begin{aligned}\Sigma_{A,(i)} &:= \left\{j : \mathbf{a}_j^T \mathbf{x}_{(i)} = b_j, j = 1, 2, \ldots, m\right\} \\ \Sigma_{x,(i)} &:= \left\{l : -x_{l,(i)} = \mathbf{e}_l^T \mathbf{x}_{(i)} = 0, l = 1, 2, \ldots, n\right\}\end{aligned} \tag{3.81}$$

i.e., it holds that

$$\begin{bmatrix} \mathbf{A} \\ -\mathbf{I}_n \end{bmatrix}_{\substack{j\in\Sigma_{A,(i)} \\ l\in\Sigma_{x,(i)}}} \mathbf{x}_{(i)} = \begin{bmatrix} \mathbf{b} \\ \mathbf{0} \end{bmatrix}_{\substack{j\in\Sigma_{A,(i)} \\ l\in\Sigma_{x,(i)}}} \tag{3.82}$$

3. Assemble matrix

$$\mathbf{Q}_{(i)} = \begin{bmatrix} \mathbf{q}_{1,(i)}^T \\ \mathbf{q}_{2,(i)}^T \\ \vdots \\ \mathbf{q}_{r,(i)}^T \end{bmatrix} := \begin{bmatrix} \mathbf{a}_j^T \\ \mathbf{e}_l^T \end{bmatrix}_{\substack{j\in\Sigma_{A,(i)} \\ l\in\Sigma_{x,(i)}}} \tag{3.83}$$

That is, rows of $\mathbf{Q}_{(i)}$ are made up of the rows of the $\mathbf{A}$ and $-\mathbf{I}_n$ matrices, for which the equality constraints hold at point $\mathbf{x}_{(i)}$.
4. Calculate the *projection vector*

$$\mathbf{p}_{(i)} = \left(\mathbf{Q}_{(i)}\mathbf{Q}_{(i)}^T\right)^{-1}\mathbf{Q}_{(i)}\mathrm{grad} f\left(\mathbf{x}_{(i)}\right) = \left[p_{1,(i)}, p_{2,(i)}, \ldots, p_{r,(i)}\right]^T \tag{3.84}$$

(a) If $r > 1$ and there exists $p_{j,(i)} < 0$ for some $j = 1, 2, \ldots, r$, adjust $\mathbf{Q}_{(i)}$ as follows

$$\mathbf{Q}_{(i)} = \begin{bmatrix} \mathbf{q}_{1,(i)}^T \\ \vdots \\ \mathbf{q}_{j-1,(i)}^T \\ \mathbf{q}_{j+1,(i)}^T \\ \vdots \\ \mathbf{q}_{r,(i)}^T \end{bmatrix} \tag{3.85}$$

I.e., the jth row is canceled from $\mathbf{Q}_{(i)}$. Set $r = r - 1$, and resolve step 4.

(b) If $r = 1$ or $p_{j,(i)} \geq 0$ for all $j = 1, 2, \ldots, r$, go to step 5.

5. Calculate the *projection matrix*

$$\mathbf{P}_{(i)} = \mathbf{I}_n - \mathbf{Q}_{(i)}^T \left(\mathbf{Q}_{(i)} \mathbf{Q}_{(i)}^T\right)^{-1} \mathbf{Q}_{(i)} \tag{3.86}$$

and the (eventual) *projection direction*

$$\mathbf{s}_{(i)} = \mathbf{P}_{(i)} \mathrm{grad} f\left(\mathbf{x}_{(i)}\right) \tag{3.87}$$

If $\mathbf{s}_{(i)} = \mathbf{0}_{n \times 1}$, set $i = i + 1$, and go to step 9.

6. The optimal (unconstrained) step length of $\mathbf{s}_{(i)}$ is calculated via the long-step method principle; i.e., by solving the optimization subproblem

$$\begin{gathered} \hat{\lambda}_{(i)}^* = \arg\max \varphi\left(\lambda_{(i)}\right) \\ \varphi\left(\lambda_{(i)}\right) := f\left(\mathbf{x}_{(i)} + \lambda_{(i)} \mathbf{s}_{(i)}\right) \end{gathered} \tag{3.88}$$

7. Calculate a provisionary updated solution estimate $\tilde{\mathbf{x}}_{(i+1)} = \mathbf{x}_{(i)} + \tilde{\lambda}_{(i)} \mathbf{s}_{(i)}$ and substitute it into (3.80); i.e., $\mathbf{x} \leftarrow \tilde{\mathbf{x}}_{(i+1)}$. Solve (3.80) for $\tilde{\lambda}_{(i)}$, leading to the admissible bounds $\tilde{\lambda}_{(i)} \in \left[\tilde{\lambda}_{(i),\min}, \tilde{\lambda}_{(i),\max}\right]$. Then,

   (a) if $\tilde{\lambda}_{(i),\min} \leq \hat{\lambda}_{(i)}^* \leq \tilde{\lambda}_{(i),\max}$, set $\lambda_{(i)}^* = \hat{\lambda}_{(i)}^*$; else,
   (b) if $\hat{\lambda}_{(i)}^* < \tilde{\lambda}_{(i),\min}$ set $\lambda_{(i)}^* = \tilde{\lambda}_{(i),\min}$; else,
   (c) if $\hat{\lambda}_{(i)}^* > \tilde{\lambda}_{(i),\max}$, set $\lambda_{(i)}^* = \tilde{\lambda}_{(i),\max}$.

8. Update the solution estimate $\mathbf{x}_{(i+1)} = \mathbf{x}_{(i)} + \lambda_{(i)}^* \mathbf{s}_{(i)}$. If $\left\|\mathbf{x}_{(i+1)} - \mathbf{x}_{(i)}\right\| \geq \varepsilon$, set $i = i + 1$ and go to step 2; else, go to step 9.
9. Stop and return $\mathbf{x}^* = \mathbf{x}_{(i+1)}$.

Note that Algorithm 3.14 can be terminated already in step 4. It can be proven that whenever no row is canceled from $\mathbf{Q}_{(i)}$, zero vector $\mathbf{s}_{(i)}$ is obtained.

A possible Matlab code for a two-variable problem and a detailed illustrative example follow.

```
% INPUTS
syms f x1 x2 lam
f = input('Enter the objective function: ');
A = input('Enter the left-hand side coefficient matrix of linear
inequality constraints (Ax<=b): ');
b = input('Enter the right-hand side value vector of linear
inequality constraints (Ax<=b): ');
x_init = input('Enter the initial point [x1 x2]: ');
e = input('Enter the desired accuracy: ');

% INIT
n = 2;
m = length(b);
```

```
A = [A; -eye(n)];
b = [b; 0; 0];
grad_f = gradient(f);
x_prev = x_init';
x = x_init';
dx = inf;

% ITERATIONS
while dx >= e
   grad_fx = subs(grad_f,[x1;x2],x);
   % DETERMINING MATRIX Q
   Q = [];
   r = 0;
   A_x = A*x;
   for i=1:1:(m+2)
      if abs(A_x(i)-b(i))<eps
         r = r+1;
         Q(r,1:n) = A(i,:);
      end
   end
   % PROJECTION VECTOR
   red = 1;
   while r>1 && red
      red = 0;
      p = (Q*Q')\Q*grad_fx;
      [min_p, i_p] = min(p);
      if min_p<0
         red = 1;
         if i_p==1
            Q = Q(2:end,:);
         elseif i_p==r
            Q = Q(1:end-1,:);
         else
            Q = [Q(1:i_p-1,:); Q(i_p+1:end,:)];
         end
         r = r-1;
      end
   end
   % PROJECTION MATRIX
   P = eye(n)-Q'*((Q*Q')\Q);
   % PROJECTION VECTOR
   s = P*grad_fx;
   if eval(norm(s))<eps
      x
      return
   end
   % ONE-DIMENSIONAL UNCONSTRAINED OPTIMIZATION
   h = subs(f,[x1;x2],x+lam*s);
   dh = diff(h,lam);
   lam0 = solve(dh==0);
   n_lam0 = length(lam0);
   lam_max=-inf;
   for j = 1:1:n_lam0
      l = 2;
```

```
      while eval(subs(diff(dh,lam),lam,lam0(j))==0)
         dh = diff(dh,lam);
         l = l+1;
      end
     if (rem(l,2)==0) && eval(subs(diff(dh,lam),lam,lam0(j))<0) &&
(lam0(j)>0)
         if lam0(j) > lam_max
            lam_max = lam0(j); % A BETTER SOLUTION
FOUND
         end
      end
   end
   % CONSTRAINTS ON LAM - SOLVING INEQUALITIES
   lam_low = -inf;
   lam_upp = inf;
   A_s = A*s;
   for i=1:1:(m+2)
      if A_s(i)>0
         lam_upp=min(lam_upp, (b(i)-A_x(i))/(A_s(i)));
      elseif A_s(i)<0
         lam_low=max(lam_low, (b(i)-A_x(i))/(A_s(i)));
      end
   end
   % DETERMINING THE EVENTUAL LAM
   if lam_max>=lam_upp
      lam_max=lam_upp;
   end
   if lam_max<=lam_low
      lam_max=lam_low;
   end
   % SOLUTION UPDATE
   dx = lam_max*s;
   x = x+dx;
   dx = norm(dx);
   x_prev = x;
end

% OUTPUT
x
```

Example 3.14 *Task*: Two work groups within one company attempt to get funds from management for their activities. In total, the management of the company can provide them with a maximum of ZAR 6 million. The companies also trade with each other through a purchase (collection) and processable raw materials. Waste containing precious metals (nickel) is processed into nickel shavings, which are repurchased by the company and used. While the first group only consumes the shavings (the discounted purchase price for the company is 1000 ZAR/1 kg), the second group creates waste, from which the processing company produces 1 ton of shavings (that does not go into storage) from every million ZAR it receives from the management. The production and storage capacity of the waste processor is 2 tons of shavings.

Find the optimal allocation of the company's cash resources between the two groups to maximize total profit within a given period, which financial analysts have determined by the following relationship:

$$f(\mathbf{x}) = f(x_1, x_2) = -2x_1^2 - x_2^2 + 16x_1 + 12x_2$$

where x_1, x_2 are financial resources dedicated to the first and second groups, respectively, in millions of ZAR. Use the gradient projection method starting from the initial estimate $\mathbf{x}_{(1)} = [0, 0]^T$ (i.e., no money is provided). Set formally $\varepsilon = 10^{-3}$ (i.e., the resolution of provided cash resources is ZAR 1000).

Solution: It is necessary to formulate mathematically the task first. The objective function (i.e., the overall company profit) is obvious. There are two constraints in the assignment. Namely, the restriction to overall allowable provided money and the production and storage capacity of the waste processing company. The former constraint reads

$$x_1 + x_2 \leq 6$$

while the latter is

$$x_1 - x_2 \leq 2$$

Moreover, it is natural that $x_1, x_2 \geq 0$, i.e., $-x_1 \leq 0, -x_2 \leq 0$. The geometrical depiction of the constraints is provided in Fig. 3.19.

Note that the analytic unconstrained maximum of the objective function $f(\mathbf{x})$ lies at $\mathbf{x}^* = [4, 6]^T$ (with the corresponding value $f(\mathbf{x}^*) = 68$) that is located outside the constraint set (see Fig. 3.19).

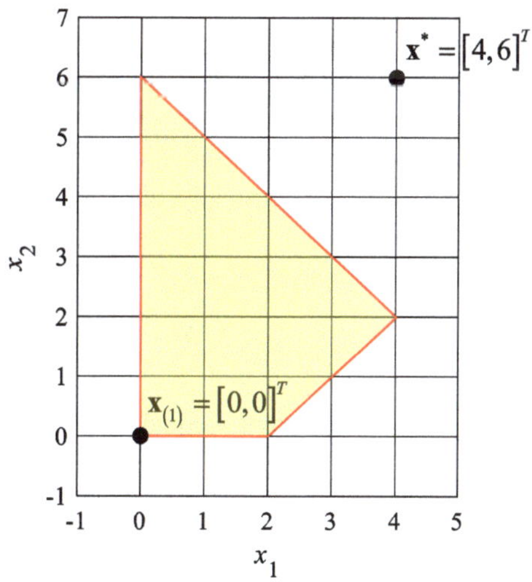

Fig. 3.19 The geometric depiction of the given constraints—Example 3.14

The gradient projection algorithm for $i = 1$ starts with determining $\mathbf{Q}_{(1)}$. The initial guess lies at the constraint set boundary, and it holds that

$$\begin{aligned} x_{1,(1)} + x_{2,(1)} &= 0 < 6 \\ x_{1,(1)} - x_{2,(1)} &= 0 < 2 \\ -x_{1,(1)} &= 0 \\ -x_{2,(1)} &= 0 \end{aligned}$$

In accordance to the notation introduced in (3.81), one has $\Sigma_{A,(1)} = \{\emptyset\}$, $\Sigma_{x,(1)} = \{1, 2\}$. Therefore,

$$\mathbf{Q}_{(1)} = \begin{bmatrix} -1 & 0 \\ 0 & -1 \end{bmatrix}$$

The gradient at $\mathbf{x}_{(1)} = [0, 0]^T$ reads $\text{grad} f\left(\mathbf{x}_{(1)}\right) = [16, 12]^T$, and the projection vector as per (3.84) is

$$\mathbf{p}_{(1)} = \left(\mathbf{Q}_{(1)}\mathbf{Q}_{(1)}^T\right)^{-1}\mathbf{Q}_{(1)}\text{grad} f\left(\mathbf{x}_{(1)}\right) = [-16, -12]^T$$

Since $p_{2,(1)} = -12 < 0$, the second row of $\mathbf{Q}_{(1)}$ is canceled. By doing this, it can be calculated that $\mathbf{p}_{(1)} = -16$. Regardless its value, there is nothing to be canceled for a scalar variable.

The corresponding projection matrix and the projection direction read

$$\mathbf{P}_{(1)} = \begin{bmatrix} 1 & 0 \\ 0 & 1 \end{bmatrix} - \mathbf{Q}_{(1)}^T\left(\mathbf{Q}_{(1)}\mathbf{Q}_{(1)}^T\right)^{-1}\mathbf{Q}_{(1)} = \begin{bmatrix} 1 & 0 \\ 0 & 0 \end{bmatrix}$$
$$\mathbf{s}_{(1)} = \mathbf{P}_{(1)}\text{grad} f\left(\mathbf{x}_{(1)}\right) = [16, 0]^T$$

respectively. Now, one can deduce from Fig. 3.19 that the updated extremum estimate $\mathbf{x}_{(2)}$ will lie at axis x_1 (i.e., following the direction $\mathbf{s}_{(1)}$ from the current estimate point $\mathbf{x}_{(1)}$) within the range $0 < x_1 \leq 2$. As the length of $\mathbf{s}_{(1)}$ equals 16, it is easy to see that the projection direction can be multiplied by $0 < \lambda_{(1)} \leq 1/8$. This graphical solution can be algebraically obtained by substituting the formula expressing the updated estimate

$$\mathbf{x}_{(2)} = \mathbf{x}_{(1)} + \lambda_{(1)}\mathbf{s}_{(1)} = \begin{bmatrix} 0 \\ 0 \end{bmatrix} + \lambda_{(1)}\begin{bmatrix} 16 \\ 0 \end{bmatrix}$$

into the linear constraint inequalities, i.e.,

$$\left.\begin{aligned} 16\lambda_{(1)} &\leq 6 \\ 16\lambda_{(1)} &\leq 2 \\ -16\lambda_{(1)} &\leq 0 \end{aligned}\right\} \Rightarrow \tilde{\lambda}_{(1),\min} = 0 \leq \lambda_{(1)} \leq 1/8 = \tilde{\lambda}_{(1),\max}$$

The particular long-step problem

$$\hat{\lambda}^*_{(1)} = \underset{\lambda_{(1)} \in R}{\arg\max}\, \varphi\left(\lambda_{(1)}\right)$$

$$\varphi\left(\lambda_{(1)}\right) := f\left(\mathbf{x}_{(1)} + \lambda_{(1)}\mathbf{s}_{(1)}\right) = f\left(\begin{bmatrix} 0 \\ 0 \end{bmatrix} + \lambda_{(1)}\begin{bmatrix} 16 \\ 0 \end{bmatrix}\right)$$
$$= 512\lambda_{(1)}\left(-\lambda_{(1)} + 0.5\right)$$

results in $\hat{\lambda}^*_{(1)} = 0.5 > \tilde{\lambda}_{(1),\max}$. Hence, the eventual step length $\lambda^*_{(1)} = \tilde{\lambda}_{(1),\max} = 1/8$ is taken, which yields $\mathbf{x}_{(2)} = [2, 0]^T$.

The second iteration step ($i = 2$) repeats the same procedure, i.e.,

$$\operatorname{grad} f\left(\mathbf{x}_{(2)}\right) = [8, 12]^T$$

$$\left.\begin{aligned} & x_{1,(2)} + x_{2,(2)} = 2 < 6 \\ & x_{1,(2)} - x_{2,(2)} = 2 \\ & -x_{1,(2)} = -2 < 0 \\ & -x_{2,(2)} = 0 \end{aligned}\right\} \Rightarrow \Sigma_{A,(2)} = \{2\},\ \Sigma_{x,(2)} = \{2\} \Rightarrow \mathbf{Q}_{(2)} = \begin{bmatrix} 1 & -1 \\ 0 & -1 \end{bmatrix}$$

$$\mathbf{p}_{(2)} = \begin{bmatrix} 8 \\ -20 \end{bmatrix}$$

After cancelling the second row in $\mathbf{Q}_{(2)}$, one gets $\mathbf{p}_{(2)} = -2$, giving rise to the projection matrix and the projection direction

$$\mathbf{P}_{(2)} = \begin{bmatrix} 0.5 & 0.5 \\ 0.5 & 0.5 \end{bmatrix}, \quad \mathbf{s}_{(2)} = \begin{bmatrix} 10 \\ 10 \end{bmatrix}$$

respectively. The eventual maximum estimate update is towards the "45° right up" direction. The admissible step length $\lambda_{(2)}$ following this direction can be calculated by substituting

$$\mathbf{x}_{(2)} + \lambda_{(2)}\mathbf{s}_{(2)} = \begin{bmatrix} 2 \\ 0 \end{bmatrix} + \lambda_{(2)}\begin{bmatrix} 10 \\ 10 \end{bmatrix}$$

into the constraint inequalities, yielding

$$\left.\begin{aligned} & 2 + 10\lambda_{(2)} + 10\lambda_{(2)} \le 6 \\ & 2 + 10\lambda_{(2)} - 10\lambda_{(2)} \le 2 \\ & -\left(2 + 10\lambda_{(2)}\right) \le 0 \\ & -10\lambda_{(2)} \le 0 \end{aligned}\right\} \Rightarrow \tilde{\lambda}_{(2),\min} = 0 \le \lambda_{(2)} \le 0.2 = \tilde{\lambda}_{(2),\max}$$

The lower bound $\tilde{\lambda}_{(2),\min} = 0$ agrees with point $[2, 0]^T$, while the upper one $\tilde{\lambda}_{(2),\max} = 0.2$ yields $[4, 2]^T$. The analytic solution (i.e., the long-step method maximizing $\varphi(\lambda_{(2)}) = f(\mathbf{x}_{(2)} + \lambda_{(2)}\mathbf{s}_{(2)})$) gives the optimal step length of $\hat{\lambda}_{(2)} = 1/3$. However, as $\hat{\lambda}_{(2)} = 1/3 > 1/5 = \tilde{\lambda}_{(2),\max}$, it is eventually set $\lambda^*_{(2)} = \tilde{\lambda}_{(2),\max} = 0.2$. Then, the updated maximum estimate reads

$$\mathbf{x}_{(3)} = \mathbf{x}_{(2)} + \lambda^*_{(2)}\mathbf{s}_{(2)} = \begin{bmatrix} 2 \\ 0 \end{bmatrix} + 0.2\begin{bmatrix} 10 \\ 10 \end{bmatrix} = \begin{bmatrix} 4 \\ 2 \end{bmatrix}$$

For $i = 3$, one can calculate that $\operatorname{grad} f(\mathbf{x}_{(3)}) = [0, 8]^T$ and

$$\left.\begin{aligned} x_{1,(3)} + x_{2,(3)} &= 6 \\ x_{1,(3)} - x_{2,(3)} &= 2 \\ -x_{1,(3)} &= -2 < 0 \\ -x_{2,(3)} &= -4 < 0 \end{aligned}\right\} \Rightarrow \Sigma_{A,(3)} = \{1, 2\},\ \Sigma_{x,(3)} = \{\emptyset\} \Rightarrow \mathbf{Q}_{(3)} = \begin{bmatrix} 1 & 1 \\ 1 & -1 \end{bmatrix}$$

$$\mathbf{p}_{(3)} = \begin{bmatrix} 4 \\ -4 \end{bmatrix}$$

Deleting the second row of $\mathbf{Q}_{(3)}$ results in $\mathbf{p}_{(3)} = 4$ and

$$\mathbf{P}_{(3)} = \begin{bmatrix} 0.5 & -0.5 \\ -0.5 & 0.5 \end{bmatrix} \Rightarrow \mathbf{s}_{(3)} = \begin{bmatrix} -4 \\ 4 \end{bmatrix}$$

Further, one can calculate that the optimal step length in the $\mathbf{s}_{(3)}$ direction equals $\hat{\lambda}_{(3)} = 1/3$ again, while the bounds are $\tilde{\lambda}_{(3)} = [0, 1]$. It means that the optimal step length is admissible, and the updated maximum estimate reads

$$\mathbf{x}_{(4)} = \mathbf{x}_{(3)} + \lambda^*_{(3)}\mathbf{s}_{(3)} = \begin{bmatrix} 4 \\ 2 \end{bmatrix} + 1/3\begin{bmatrix} -4 \\ 4 \end{bmatrix} = \begin{bmatrix} 8/3 \\ 10/3 \end{bmatrix}$$

This iteration step is outlined in Fig. 3.20. (Note that the vector lengths are reduced in the figure.)

Further, it holds at $\mathbf{x}_{(4)}$ that

$$\operatorname{grad} f(\mathbf{x}_{(4)}) = [16/3, 16/3]^T,\ \mathbf{Q}_{(4)} = \begin{bmatrix} 1 & 1 \end{bmatrix},\ \mathbf{p}_{(4)} = 4$$

(i.e., nothing is canceled in $\mathbf{Q}_{(4)}$) and

$$\mathbf{P}_{(4)} = \begin{bmatrix} 0.5 & -0.5 \\ -0.5 & 0.5 \end{bmatrix} \Rightarrow \mathbf{s}_{(4)} = \begin{bmatrix} 0 \\ 0 \end{bmatrix}$$

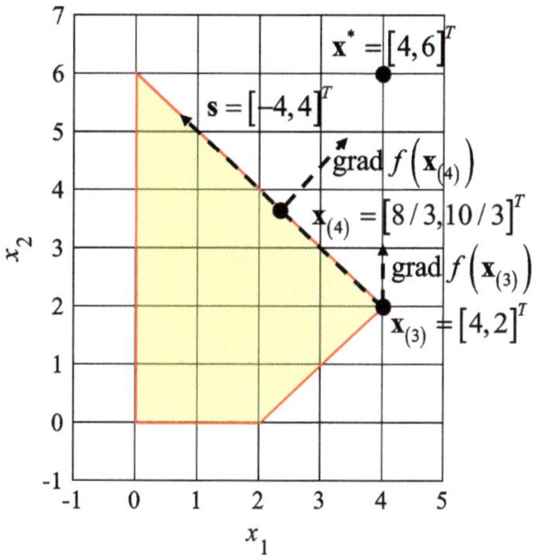

Fig. 3.20 The outline of the 3rd iteration step—Example 3.14

It means that $\mathbf{x}^* = \mathbf{x}_{(4)} = [8/3,\ 10/3]^T$ since the projection direction is the zero vector. It is natural, as $\mathrm{grad} f\left(\mathbf{x}_{(4)}\right)$ is orthogonal to the constraint boundary (see Fig. 3.20).

Note that $f(\mathbf{x}^*) = 57.\overline{3}$; i.e., the total profit of the company will be ZAR 57,333,000, when the subvention of ZAR 6 million will be optimally distributed among the two groups by the ratio $8/10 = 4/5$.

Example 3.14 demonstrates that the gradient projection method does not provide the minimum number of iteration steps in general. Indeed, it is clear from Fig. 3.19 that the minimum number of steps would follow this solution: $\mathbf{x}_{(1)} = [0,\ 0]^T$, $\mathbf{x}_{(2)} = [0,\ 6]^T$, $\mathbf{x}_{(3)} = \mathbf{x}^* = [8/3,\ 10/3]^T$.

3.3.2 *Penalty Function*

Since the gradient projection method is relatively computationally complex, the question arises as to whether it is possible to directly use some previously known iterative method, whether gradient or comparative. One of the possibilities is using the so-called penalty function, which makes it possible to include constraints directly in the objective function; i.e., to convert the task into an unconstrained problem.

This principle can be sufficiently applied to convex-set optimization tasks, for which the extremum is supposed to lie at the constraint boundary.

Consider the following constrained optimization problem

$$
\begin{aligned}
\min f(\mathbf{x}) &\in \mathbb{R}, \mathbf{x} \in \mathbb{R}^n \\
g_j(\mathbf{x}) &\leq 0,\ j = 1, \ldots, m_g \\
h_l(\mathbf{x}) &= 0,\ l = 1, \ldots, m_h
\end{aligned}
\tag{3.89}
$$

that can be transformed into the unconstrained task

$$
\begin{aligned}
&\min p(\mathbf{x}) \in \mathbb{R}, \mathbf{x} \in \mathbb{R}^n \\
&p(\mathbf{x}) := f(\mathbf{x}) + \alpha \left(\sum_{j=1}^{m_g} \left(\max\left(0, g_j(\mathbf{x})\right)\right)^k + \sum_{l=1}^{m_h} |h_l(\mathbf{x})|^k \right) \\
&k \in \mathbb{N} \geq 1,\ \ \alpha \in \mathbb{R} > 0
\end{aligned}
\tag{3.90}
$$

The principle of using the penalty function is that if $\mathbf{x}$ satisfies the constraints, it holds that $p(\mathbf{x}) = f(\mathbf{x})$, otherwise $p(\mathbf{x}) > f(\mathbf{x})$, and then the optimization procedure for solving the unconstrained extremum attempts to suppress the difference $p(\mathbf{x}) - f(\mathbf{x})$ and thus obtain an admissible $\mathbf{x}$. The penalty parameter α can be constant or it may vary in each iteration step. The following theorem applies.

Theorem 3.1 *Let $f(\mathbf{x})$, $g_j(\mathbf{x})$, $h_l(\mathbf{x})$, for $j = 1, \ldots, m_g$, $l = 1, \ldots, m_h$, be continuous and, moreover, $f(\mathbf{x})$ be bounded from below. Let there exists an ascending sequence of positive real numbers $\{\alpha_i\}_{i=1}^{\infty}$, $\lim_{i\to\infty} \alpha_i = \infty$. Then, it holds that*

$$
\begin{aligned}
&\lim_{i\to\infty} p\left(\mathbf{x}_{(i)}, \alpha_{(i)}\right) = f\left(\mathbf{x}^*\right) \\
&\lim_{i\to\infty} \left(\sum_{j=1}^{m_g} \left(\max\left(0, g_j\left(\mathbf{x}_{(i)}\right)\right)\right)^k + \sum_{l=1}^{m_h} \left|h_l\left(\mathbf{x}_{(i)}\right)\right|^k \right) = 0
\end{aligned}
\tag{3.91}
$$

where subscript (i) expresses the iteration step number.

Theorem 3.1 means that if $\{\alpha_i\}_{i=1}^{\infty}$, $\lim\limits_{i\to\infty} \alpha_i = \infty$ is introduced, the extremum $\mathbf{x}^*$ lying at the constraint boundary is asymptotically reached.

The use of the penalty function can be summarized by the following algorithm.

Algorithm 3.15 (Penalty function principle)

1. Consider minimization problem (3.89) and transform it into (3.90) with fitness function $p(\mathbf{x}, \alpha)$. Construct an ascending sequence of positive real numbers $\{\alpha_i\}_{i=1}^{\infty}$, $\lim_{i\to\infty} \alpha_i = \infty$. Select the desired accuracy $\varepsilon > 0$. Set $i = 1$ and $\mathbf{x}_{(i)} = \mathbf{x}_{(1)}$.
2. Solve the unconstrained optimization problem (3.90) with $\alpha_{(i)} = \alpha_i$ using any iterative method from Sects. 3.1.2 or 3.2.2, giving rise to the solution estimate $\mathbf{x}_{(i+1)}$.

3. If $\left\|\mathbf{x}_{(i+1)} - \mathbf{x}_{(i)}\right\| \geq \varepsilon$, set $i = i + 1$ and go to step 2; else, go to step 4.
4. Stop and return $\mathbf{x}^* = \mathbf{x}_{(i+1)}$.

The advantage of the method is apparently the possibility of solving the unconstrained extremum problem instead of the constrained one. However, the penalty function principle is unsuitable for solving nonconvex problems, and for high α_i, the Hessian of $p\left(\mathbf{x}_{(i)}, \alpha_{(i)}\right)$ may become ill-conditioned (singular).

An illustrative numerical example follows.

Example 3.15 *Task*: Consider the optimization task formulated in Example 3.14. Solve it by applying the penalty function principle for $\alpha_{(i)} = 10^i$ using the Nelder–Mead method.

Solution: Introduce $p(\mathbf{x}, \alpha)$ as

$$\begin{aligned} p(\mathbf{x}, \alpha) &= \tilde{f}(\mathbf{x}) + \alpha\left(\sum_{j=1}^{4}\left(\max\left(0, g_j(\mathbf{x})\right)\right)^2\right) \\ &= 2x_1^2 + x_2^2 - 16x_1 - 12x_2 \\ &\quad + \alpha\begin{pmatrix} (\max(0, x_1 + x_2 - 6))^2 + (\max(0, x_1 - x_2 - 2))^2 \\ +(\max(0, -x_1))^2 + (\max(0, -x_2))^2 \end{pmatrix} \end{aligned}$$

where $\tilde{f}(\mathbf{x}) = -f(\mathbf{x}) = 2x_1^2 + x_2^2 - 16x_1 - 12x_2$ since the given optimization problem is formulated for searching for a maximum, while Algorithm 3.15 solves a minimum, as seen in (2.22).

In each iteration step, the goal is to solve the subproblem $\mathbf{x}_{(i)}^* = \arg\min p\left(\mathbf{x}_{(i)}, \alpha_{(i)}\right)$ via the Nelder–Mead method (see Sect. 3.1.2.2) with the setting as in Example 3.5. The corresponding numerical results are summarized in Table 3.9.

The iterated solutions displayed in Table 3.9 apparently converge to the analytic solution $\mathbf{x}^* = [8/3,\ 10/3]^T = \left[2.\overline{6},\ 3.\overline{3}\right]^T$, as indicated in Theorem 3.1. Note that $\mathbf{x}_{(14)}^l = [2.434, 3.520]^T$ with $p\left(\mathbf{x}_{(14)}^l\right) = -f\left(\mathbf{x}_{(14)}^l\right) = -56.945$, where the simplex midpoint reads $\mathbf{x}_{(14)}^{mid} = [2.309, 3.592]^T$, giving $p\left(\mathbf{x}_{(14)}^{mid}\right) = -f\left(\mathbf{x}_{(14)}^{mid}\right) = -56.481$.

The iterative evolution of $\mathbf{x}_{(i)}^l$ (i.e., "best" simplex vertices) is shown in Fig. 3.21.

3.3.3 Barrier Function

The use of the barrier function represents another approach to incorporating the constraints into the objective function. However, its nature differs from the penalty function in principle. Whereas the penalty function returns its positive values whenever the constraints are broken, the barrier one does not allow to cross the border.

Table 3.9 The use of the penalty function—Nelder–Mead method—Example 3.15

i	$\mathbf{x}^1_{(i)}$	$\mathbf{x}^2_{(i)}$	$\mathbf{x}^3_{(i)}$	$\mathbf{x}^{p0}_{(i)}$	$p\left(\mathbf{x}^l_{(i)}\right)$	$f\left(\mathbf{x}^l_{(i)}\right)$
	$p\left(\mathbf{x}^1_{(i)}\right)$	$p\left(\mathbf{x}^2_{(i)}\right)$	$p\left(\mathbf{x}^3_{(i)}\right)$	$p\left(\mathbf{x}^{p0}_{(i)}\right)$	$\alpha_{(i)}$	Operation
1	$[0, 0]^T$	$[1, 0]^T$	$[0, 1]^T$	$[1.5, 1.5]^T$	– 14	14
	0	– 14	– 11	– 35.25	10	E
2	$[1.5, 1.5]^T$	$[1, 0]^T$	$[0, 1]^T$	$[2.5, 0.5]^T$	– 35.25	35.25
	– 35.25	– 14	– 11	– 33.25	100	0
3	$[1.5, 1.5]^T$	$[1, 0]^T$	$[2.5, 0.5]^T$	$[4, 3]^T$	– 35.25	35.25
	– 35.25	– 14	– 33.25	941	1000	E
4	$[1.5, 1.5]^T$	$[3, 2]^T$	$[2.5, 0.5]^T$	$[1.75, 4.25]^T$	– 50	50
	– 35.25	– 50	– 33.25	– 54.81	10^4	E
5	$[1.5, 1.5]^T$	$[3, 2]^T$	$[1.75, 4.25]^T$	$[1.937, 2.2313]^T$	– 54.81	54.81
	– 35.25	– 50	– 54.81	– 45.89	10^5	IC
6	$[1.937, 2.2313]^T$	$[3, 2]^T$	$[1.75, 4.25]^T$	$[2.156, 2.719]^T$	– 54.81	54.81
	– 45.89	– 50	– 54.81	– 50.43	10^6	IC
7	$[2.156, 2.719]^T$	$[3, 2]^T$	$[1.75, 4.25]^T$	$[2.477, 2.742]^T$	– 54.81	54.81
	– 50.43	– 50	– 54.81	– 52.74	10^7	IC
8	$[2.156, 2.719]^T$	$[2.477, 2.742]^T$	$[1.75, 4.25]^T$	$[2.135, 3.107]^T$	– 54.81	54.81
	– 50.43	– 52.74	– 54.81	– 52.67	10^8	IC
9	$[2.135, 3.107]^T$	$[2.477, 2.742]^T$	$[1.75, 4.25]^T$	$[2.092, 3.885]^T$	– 54.81	54.81
	– 52.67	– 52.74	– 54.81	– 56.24	10^9	E
10	$[2.092, 3.885]^T$	$[2.477, 2.742]^T$	$[1.75, 4.25]^T$	$[2.199, 3.405]^T$	– 56.24	56.24
	– 56.24	– 52.74	– 54.81	– 54.78	10^{10}	IC
11	$[2.092, 3.885]^T$	$[2.477, 2.742]^T$	$[1.75, 4.25]^T$	$[2.060, 3.736]^T$	– 56.24	56.24
	– 56.24	– 52.74	– 54.81	– 55.35	10^{11}	IC
12	$[2.092, 3.885]^T$	$[2.060, 3.736]^T$	$[1.75, 4.25]^T$	$[2.402, 3.371]^T$	– 56.24	56.24
	– 56.24	– 55.35	– 54.81	– 55.98	10^{12}	0
13	$[2.092, 3.885]^T$	$[2.060, 3.736]^T$	$[2.402, 3.371]^T$	$[2.434, 3.520]^T$	– 56.24	56.24
	– 56.24	– 55.35	– 55.98	– 56.94	10^{13}	E

Note that the "operation" was defined in Example 3.5.

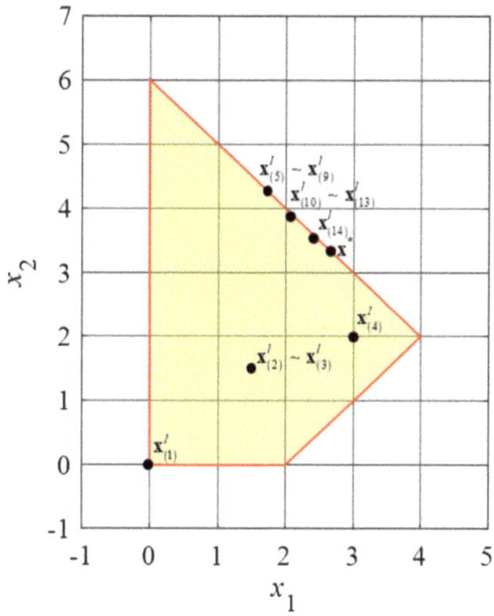

Fig. 3.21 Iterations of the "best" vertices $\mathbf{x}^l_{(i)}$—Example 3.15

If a solution satisfies the given constraints and is "far" from the border, the barrier function returns low non-negative (or even negative) values. Contrariwise, whenever a particular solution approaches the border, the barrier function value starts rising sharply.

The barrier function is reasonable to be used for convex-set optimization problems, for which the extremum is expected to *not* be at the constraint boundary.

Consider the following constrained optimization problem

$$\begin{aligned} &\min f(\mathbf{x}) \in \mathbb{R}, \mathbf{x} \in \mathbb{R}^n \\ &g_j(\mathbf{x}) \leq 0,\ j = 1, \ldots, m_g \end{aligned} \tag{3.92}$$

that can be transformed to the unconstrained task

$$\begin{aligned} &\min p(\mathbf{x}) \in \mathbb{R}, \mathbf{x} \in \mathbb{R}^n \\ &p(\mathbf{x}) := f(\mathbf{x}) + \alpha \sum_{j=1}^{m} h\big(g_j(\mathbf{x})\big) \\ &\lim_{g_j(\mathbf{x}) \to 0^-} h\big(g_j(\mathbf{x})\big) = \infty,\ j = 1, \ldots, m;\ \ \alpha \in \mathbb{R} > 0 \end{aligned} \tag{3.93}$$

Hence, solution $\mathbf{x}^* = \arg\min p(\mathbf{x})$ is to be then found. Again, any (iterative) method can be used for searching an unconstrained extremum; however, the initial guess $\mathbf{x}_{(1)}$ must satisfy

$$g_j\big(\mathbf{x}_{(1)}\big) \leq 0, \forall j \tag{3.94}$$

For example, functions $h\big(g_j\big)$ can be selected as

$$h(g_j(\mathbf{x})) = \sum_{j=1}^{m} -\frac{1}{g_j(\mathbf{x})}$$
$$h(g_j(\mathbf{x})) = \sum_{j=1}^{m} -\ln(-g_j(\mathbf{x})) \tag{3.95}$$

Notice that the above-introduced principle cannot be used for equality constraints as it would lead to $p(\mathbf{x}) \to \infty$.

As the iteration methods are discontinuous in their nature, a solution estimate can appear in a non-admissible set (region). Such a subsolution then be canceled or its fitness function value can technically be set as infinity (e.g., at one of the simplex vertices).

The barrier function principle algorithm can be summarized as follows.

Algorithm 3.16 (Barrier function principle)

1. Consider minimization problem (3.92). Select $\alpha > 0$ and $h(g_j)$, and transform the problem into (3.93) with fitness function $p(\mathbf{x}, \alpha)$. Select the desired accuracy $\varepsilon > 0$. Set $i = 1$ and select $\mathbf{x}_{(i)} = \mathbf{x}_{(1)}$ satisfying (3.94).
2. Solve the unconstrained optimization problem (3.93) using any iterative method from Sects. 3.1.2 or 3.2.2, giving rise to the solution estimate $\mathbf{x}_{(i+1)}$.
3. If $\left\|\mathbf{x}_{(i+1)} - \mathbf{x}_{(i)}\right\| \geq \varepsilon$, set $i = i + 1$ and go to step 2; else, go to step 4.
4. Stop and return $\mathbf{x}^* = \mathbf{x}_{(i+1)}$.

The following example demonstrates an attempt to solve the task defined in Example 3.14.

Example 3.16 *Task*: Consider the optimization task formulated in Example 3.14.

Solve it by applying the barrier function principle for $\alpha = 10$ using the Nelder–Mead method.

Solution: Construct $p(\mathbf{x}, \alpha)$ as

$$\begin{aligned} p(\mathbf{x}) &= \tilde{f}(\mathbf{x}) + \alpha\left(\sum_{j=1}^{4} -\ln(-g_j(\mathbf{x}))\right) \\ &= 2x_1^2 + x_2^2 - 16x_1 - 12x_2 \\ &\quad + \alpha(-\ln(-x_1 - x_2 + 6) - \ln(-x_1 + x_2 + 2) - \ln(x_1) - \ln(x_2)) \end{aligned}$$

where $\tilde{f}(\mathbf{x}) = -f(\mathbf{x}) = 2x_1^2 + x_2^2 - 16x_1 - 12x_2$ again.

Select, e.g., $\mathbf{x}_{(1)} = [0.1, 0.1]^T$ and the initial simple edge length $\Delta = 1$. The iterative solution results are summarized in Table 3.10.

After the number of 13 iteration steps, it is evident that the procedure does not converge to the actual extremum, but the solution estimates approach another point inside the constraint set (see the evolution of $\mathbf{x}_{(i)}^l$ displayed in Fig. 3.22).

Table 3.10 The use of the barrier function—Nelder–Mead method—Example 3.16

i	$\mathbf{x}^1_{(i)}$	$\mathbf{x}^2_{(i)}$	$\mathbf{x}^3_{(i)}$	$\mathbf{x}^{p0}_{(i)}$	$p\left(\mathbf{x}^l_{(i)}\right)$	$f\left(\mathbf{x}^l_{(i)}\right)$
	$p\left(\mathbf{x}^1_{(i)}\right)$	$p\left(\mathbf{x}^2_{(i)}\right)$	$p\left(\mathbf{x}^3_{(i)}\right)$	$p\left(\mathbf{x}^{p0}_{(i)}\right)$		Operation
1	$[0.1, 0.1]^T$	$[1.1, 0.1]^T$	$[0.1, 1.1]^T$	$[1.6, 1.6]^T$	– 18.17	13.57
	18.77	– 9.98	– 18.17	– 63.75		E
2	$[1.6, 1.6]^T$	$[1.1, 0.1]^T$	$[0.1, 1.1]^T$	$[0.6, 2.6]^T$	– 63.75	37.12
	– 63.75	– 9.98	– 18.17	– 61.93		0
3	$[1.6, 1.6]^T$	$[0.6, 2.6]^T$	$[0.1, 1.1]^T$	$[2.1, 3.1]^T$	– 63.75	37.12
	– 63.75	– 61.93	– 18.17	– 79.86		E
4	$[1.6, 1.6]^T$	$[0.6, 2.6]^T$	$[2.1, 3.1]^T$	$[3.1, 2.1]^T$	– 79.86	52.37
	– 63.75	– 61.93	– 79.86	– 67.67		0
5	$[1.6, 1.6]^T$	$[3.1, 2.1]^T$	$[2.1, 3.1]^T$	$[2.1, 2.1]^T$	– 79.86	52.37
	– 63.75	– 67.67	– 79.86	– 73.22		IC
6	$[2.1, 2.1]^T$	$[3.1, 2.1]^T$	$[2.1, 3.1]^T$	$[1.1, 3.1]^T$	– 79.86	52.37
	– 73.22	– 67.67	– 79.86	– 74.78		0
7	$[2.1, 2.1]^T$	$[1.1, 2.1]^T$	$[2.1, 3.1]^T$	$[1.1, 4.1]^T$	– 79.86	52.37
	– 73.22	– 74.78	– 79.86	– 76.50		0
8	$[1.1, 4.1]^T$	$[1.1, 2.1]^T$	$[2.1, 3.1]^T$	$[1.35, 3.35]^T$	– 79.86	52.37
	– 76.50	– 74.78	– 79.86	– 79.27		IC
9	$[1.1, 4.1]^T$	$[1.35, 3.35]^T$	$[2.1, 3.1]^T$	$[1.413, 3.663]^T$	– 79.86	52.37
	– 76.50	– 78.51	– 79.86	– 56.24		IC
10	$[1.413, 3.663]^T$	$[1.35, 3.35]^T$	$[2.1, 3.1]^T$	$[1.553, 3.366]^T$	– 79.86	52.37
	– 56.24	– 78.51	– 79.86	– 79.79		IC
11	$[1.413, 3.663]^T$	$[1.553, 3.366]^T$	$[2.1, 3.1]^T$	$[1.620, 3.448]^T$	– 79.86	52.37
	– 56.24	– 79.79	– 79.86	– 80.08		IC
12	$[1.620, 3.448]^T$	$[1.553, 3.366]^T$	$[2.1, 3.1]^T$	$[1.706, 3.320]^T$	– 80.08	50.15
	– 80.08	– 79.79	– 79.86	– 80.22		IC
13	$[1.620, 3.448]^T$	$[1.706, 3.320]^T$	$[2.1, 3.1]^T$	$[1.882, 3.242]^T$	– 80.22	50.30
	– 80.08	– 80.22	– 79.86	– 80.30		IC

Another interesting feature is a deterioration of $f\left(\mathbf{x}^l_{(i)}\right)$ during the iteration computation while the values of $p\left(\mathbf{x}^l_{(i)}\right)$ are constantly decreasing (see $i = 12$ in Table 3.10).

For completeness, the eventual solution estimates after $i = 14$ are $\mathbf{x}^l_{(14)} = [1.882, 3.242]^T$ with $p\left(\mathbf{x}^l_{(14)}\right) = -80.22$, $f\left(\mathbf{x}^l_{(14)}\right) = 50.3$, where the simplex midpoint reads $\mathbf{x}^{mid}_{(14)} = [1.736, 3.336]^T$, giving $p\left(\mathbf{x}^{mid}_{(14)}\right) = -80.277$ and $f\left(\mathbf{x}^{mid}_{(14)}\right) = 50.652$.

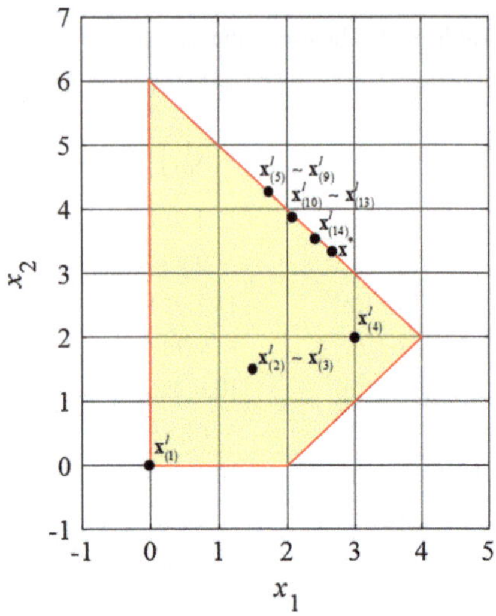

Fig. 3.22 Iterations of the "best" vertices $\mathbf{x}^l_{(i)}$—Example 3.16

3.4 Random Search Methods

These computational methods are based on searching the solution space with some combination of probability and heuristics. Technically, they should be included in the family of gradient-free methods (see Sect. 3.1); however, a separate chapter is devoted to them due to their specificity. The principle of a "reasonable" option to choose in the next iterative step is often applied based on the evaluation of the current (or also previous) step(s) result(s).

They can be classified, e.g., as follows:

- **Simple** search
- **Adaptive** search
  - with a **backstep**
  - with a **linear recalculation**
  - with a **prediction**
  - with a **penalization**
- Metaheuristics
  - Evolutionary algorithms (genetic programming, evolutionary programming, differential evolution, etc.)
  - Swarm algorithms (particle swarm optimization, ant colony optimization, bees algorithm, etc.)
  - Simulated annealing
  - Local search methods (iterated local search, stochastic local search).

Note that there can be identified many other modern, complex, and sophisticated methods that use artificial intelligence principles. However, they go beyond this book, and readers are referred to particular literature resources for further details. As this book intends to present only standard and well-established essential principles, basic ideas of the simple search and adaptive search methods are introduced.

A general principle of these methods can be expressed by the iterative formula

$$\begin{aligned}\mathbf{x}_{(i+1)} &= \mathbf{x}_{(i)} + \Delta\mathbf{x}_{(i)}\\ \Delta\mathbf{x}_{(i)} &= a_{(i)}\boldsymbol{\xi}_{(i)}\varphi\left(i, \Delta f\left(\mathbf{x}_{(i-1)}\right), \Delta\mathbf{x}_{(i-1)}\right)\end{aligned} \tag{3.96}$$

where

$$\begin{aligned} a_{(i)} = a_k,\ \{a_k\}_{k=1}^{\infty} &: \sum_{k=1}^{\infty} a_k \to \infty, \sum_{k=1}^{\infty} a_k^2 < \infty\\ \Delta f\left(\mathbf{x}_{(i-1)}\right) &= f\left(\mathbf{x}_{(i)}\right) - f\left(\mathbf{x}_{(i-1)}\right)\end{aligned} \tag{3.97}$$

and $\varphi(\cdot)$ represents a function evaluating the preceding step, $\boldsymbol{\xi} \in \mathbb{R}^n$ expresses a vector of uniformly distributed random variables within the range [0, 1] (or another suitable interval), and i denotes the iteration step. Sequence $\{a_k\}_{k=1}^{\infty}$ can, e.g., be composed of its elements

$$\begin{aligned} a_k &= \frac{1}{k}\\ a_k &= \frac{b}{k}, b > 0\\ a_k &= \frac{b}{k+c}, b > 0, c > 0\end{aligned} \tag{3.98}$$

The particular methods principally differ in computing $\Delta\mathbf{x}_{(i)}$. They either penalize the unsuccessful step or appraise the successful one.

Below, an unconstrained *minimization* problem is considered. Whenever the task is to find a maximum of $f(\mathbf{x})$, recall that one can simply minimize the fitness function $h(\mathbf{x}) = -f(\mathbf{x})$ instead (see Sect. 2.4.1).

3.4.1 Simple Search

This method has also attributes "learningless" or "without evaluation". It does not reflect the result of the last iteration step, and it can be expressed by the very simple solution update

$$\Delta\mathbf{x}_{(i)} = a_{(i)}\boldsymbol{\xi}_{(i)} \tag{3.99}$$

3.4.2 Adaptive Search with a Backstep

The principle of this method is that if the $(i-1)$th step was successful, the ith step is generated by a simple random search method; otherwise, the extremum (minimum) estimate goes back to $\mathbf{x}_{(i-1)}$, i.e.

$$\Delta\mathbf{x}_{(i)} = 0.5\left(1 - \mathrm{sgn}\Delta f\left(\mathbf{x}_{(i-1)}\right)\right)a_{(i)}\boldsymbol{\xi}_{(i)} - 0.5\left(1 + \mathrm{sgn}\Delta f\left(\mathbf{x}_{(i-1)}\right)\right)\Delta\mathbf{x}_{(i-1)} \quad (3.100)$$

where sgn means the signum function, i.e., $\mathrm{sgn}t = 1$ if $t > 0$, $\mathrm{sgn}t = -1$ if $t < 0$ (the technical case $\mathrm{sgn}0 = 0$ can be omitted in practice). An alternative definition is $\mathrm{sgn}t = t/|t|$. Formula (3.100) can also be expressed by a Matlab code as follows (assuming that $\Delta\mathbf{x}_{(i-1)} = \mathbf{x}_{(i)} - \mathbf{x}_{(i-1)}$, $\Delta f\left(\mathbf{x}_{(i-1)}\right) = f\left(\mathbf{x}_{(i)}\right) - f\left(\mathbf{x}_{(i-1)}\right)$, and $a_{(i)} = a_k$ are known).

```
% dx_i1=x_i-x_i1, df_i1=f_i-f_i1, a_i are known
if sign(df_i1)
   dx_i = -dx_i1;
else
   ksi = rand;
   dx_i = a_i*ksi;
end
```

3.4.3 Adaptive Search with a Linear Recalculation

The idea is similar to the search with a backstep; however, the solution estimate does not return to the preceding point $\mathbf{x}_{(i-1)}$, in case of failure, but towards the opposite direction symmetrically to $\mathbf{x}_{(i-1)}$. The update rule reads

$$\Delta\mathbf{x}_{(i)} = 0.5\left(1 - \mathrm{sgn}\Delta f\left(\mathbf{x}_{(i-1)}\right)\right)a_{(i)}\boldsymbol{\xi}_{(i)} - \left(1 + \mathrm{sgn}\Delta f\left(\mathbf{x}_{(i-1)}\right)\right)\Delta\mathbf{x}_{(i-1)} \quad (3.101)$$

that can also be formulated as

```
% dx_i1=x_i-x_i1, df_i1=f_i-f_i1, a_i are known
if sign(df_i1)
   dx_i = -2*dx_i1;
else
   ksi = rand;
   dx_i = a_i*ksi;
end
```

3.4.4 Adaptive Search with a Prediction

The idea of this method is opposed to the preceding two methods. It does not punish the failure but appraises the successful step. Namely, if the step is successful, the solution estimation update is searched in the direction $\mathbf{x}_{(i)} - \mathbf{x}_{(i-1)}$, symmetrically to $\mathbf{x}_{(i)}$. Otherwise, the new direction is given randomly by the simple search principle.

The update formula is

$$\Delta\mathbf{x}_{(i)} = 0.5\big(1 - \mathrm{sgn}\Delta f\left(\mathbf{x}_{(i-1)}\right)\big)\Delta\mathbf{x}_{(i-1)} + 0.5\big(1 + \mathrm{sgn}\Delta f\left(\mathbf{x}_{(i-1)}\right)\big)a_{(i)}\boldsymbol{\xi}_{(i)} \quad (3.102)$$

that can alternatively be formulated by the decision conditions

```
% dx_i1=x_i-x_i1, df_i1=f_i-f_i1, a_i are known
if sign(df_i1)
   ksi = rand;
   dx_i = a_i*ksi;
else
   dx_i = dx_i1;
end
```

3.4.5 Adaptive Search with a Penalization

In this case, the methodology is similar to the search with a prediction; however, the direction of the random search in case of failure is generated differently:

$$\begin{aligned}\Delta\mathbf{x}_{(i)} &= 0.5\big(1 - \mathrm{sgn}\Delta f\left(\mathbf{x}_{(i-1)}\right)\big)\Delta\mathbf{x}_{(i-1)}\\ &\quad + 0.5\big(1 + \mathrm{sgn}\Delta f\left(\mathbf{x}_{(i-1)}\right)\big)a_{(i)}\boldsymbol{\xi}_{(i)} \odot \mathrm{sgn}\Delta\mathbf{x}_{(i-1)}\end{aligned} \quad (3.103)$$

where $\odot$ means the Hadamard (i.e., element-wise) product.

Note that (3.103) can also be expressed via the following if–then statement.

```
% dx_i1=x_i-x_i1, df_i1=f_i-f_i1, a_i are known
if sign(df_i1)
   ksi = rand;
   dx_i = a_i*ksi.* sign(x_i);
else
   dx_i = dx_i1;
end
```

The presented simple adaptive random searching methods are easy to implement. It was proven that these algorithms converge to the extremum of a convex function when considering (theoretically) infinitely many iterative steps. However, the obtained numerical solution is often protracted.

An example of attempting to use the above-given random search methods follows.

Example 3.17 *Task*: Consider the optimization task formulated in Example 3.14. Solve it by applying the penalty function with $\alpha_{(i)} = 10^i$ (see Example 3.15) using

(a) the simple search method
(b) the adaptive search method with a backstep
(c) the adaptive search method with a linear recalculation
(d) the adaptive search method with a prediction
(e) the adaptive search method with a penalization.

Let the initial guess be $\mathbf{x}_{(1)} = [1, 1]^T$, $a_{(i)} = 1/i$, and $\boldsymbol{\xi}$ have entries $\xi_j \in [-1, 1]$ (i.e., the initial solution update does not leave the given constraints; see, e.g., Fig. 3.19). Compare the results with those of Example 3.15 (where the Nelder–Mead method was used).

Solution: The penalty function $p(\mathbf{x}, \alpha)$ was introduced in Example 3.14. Recall that the original maximization problem of $f(\mathbf{x})$ is transposed to the minimization of $p(\mathbf{x}, \alpha)$. For each method, 20 runs of a particular algorithm have been performed and the best result has eventually been selected. Readers are provided with the results of the first two steps of each method, the corresponding data table, and concise comments on observations of the bunch of the 20 runs.

One can easily observe that $p\left(\mathbf{x}_{(1)}, 10\right) = -f\left(\mathbf{x}_{(1)}\right) = -25$.

(a) The simple random search method gives inconsistent results with the largest variance. Some algorithm runs return relatively good solution estimations; however, others might be very poor or out of the constraints. It is due to a lack of "intelligence" and adaptiveness of the method, and only random "zig-zag" movements in the solution space are made.

The best solution (out of 20 runs) follows. It was computed that

$$a_{(1)} = 1, \ \boldsymbol{\xi}_{(1)} = \begin{bmatrix} 0.8547 \\ 0.8350 \end{bmatrix} \Rightarrow \Delta\mathbf{x}_{(1)} = a_{(1)}\boldsymbol{\xi}_{(1)} = \begin{bmatrix} 0.8547 \\ 0.8350 \end{bmatrix} \Rightarrow \mathbf{x}_{(2)} = \begin{bmatrix} 1.8547 \\ 1.8350 \end{bmatrix}$$

with $p\left(\mathbf{x}_{(2)}, 100\right) = -f\left(\mathbf{x}_{(2)}\right) = -41.4482$, i.e., $\operatorname{sgn}\Delta p\left(\mathbf{x}_{(2)}, 100\right) = -1$ that represents a successful step. However, it does not have an effect to the next step where $\Delta\mathbf{x}_{(2)} = 0.5\boldsymbol{\xi}_{(1)} = 0.5\begin{bmatrix} 0.4271 \\ 0.2367 \end{bmatrix} = \begin{bmatrix} 0.2136 \\ 0.1183 \end{bmatrix}$ is generated randomly, etc. Table 3.11 shows the complete result data.

Although it is not obvious from Table 3.11, a norm of $\Delta\mathbf{x}_{(i)}$ is asymptotically (yet not uniformly) decreasing. Notice also that the random search methods do not guarantee the monotonicity of the solution estimate function values (unlike, e.g., the simplex methods).

(b) It has been observed from the performed numerical tests that the adaptive search method with a backstep provides a bit more consistent result (i.e., with a lower variability). However, as it is a defensive (safe) method, the results usually

Table 3.11 The use of the simple search method (with a penalty function)—Example 3.17

i	α / $\Delta \mathbf{x}_{(i-1)}$	a / $\Delta \mathbf{x}_{(i)}$	$\boldsymbol{\xi}$ / $p(\mathbf{x}_{(i)})$	$\mathbf{x}_{(i)}$ / $f(\mathbf{x}_{(i)})$	sgn $\Delta f(\mathbf{x}_{(i-1)})$
1	10	1	$[0.8547, 0.8350]^T$	$[1, 1]^T$	–
	–	$[0.8547, 0.3350]^T$	-25	25	
2	100	0.5	$[0.4271, 0.2367]^T$	$[1.8547, 1.8350]^T$	-1
	$[0.8547, 0.8350]^T$	$[0.2136, 0.1183]^T$	-41.4482	41.4482	
3	1000	1/3	$[-0.3134, 0.8721]^T$	$[2.0683, 1.9533]^T$	-1
	$[0.2136, 0.1183]^T$	$[-0.1045, 0.2907]^T$	-44.1614	44.1614	
4	10^4	1/4	$[-0.7505, 0.4612]^T$	$[1.9638, 2.2440]^T$	-1
	$[-0.1045, 0.2907]^T$	$[-0.1876, 0.1153]^T$	-45.6004	45.6004	
5	10^5	1/5	$[0.2930, 0.6663]^T$	$[1.7762, 2.3593]^T$	1
	$[-0.1876, 0.1153]^T$	$[0.0586, 0.1333]^T$	-44.8547	44.8547	
6	10^6	1/6	$[-0.2034, 0.4996]^T$	$[1.8348, 2.4926]^T$	-1
	$[0.0586, 0.1333]^T$	$[-0.0339, 0.0833]^T$	-46.3216	46.3216	
7	10^7	1/7	$[0.6704, -0.3551]^T$	$[1.8009, 2.5758]^T$	-1
	$[-0.0339, 0.0833]^T$	$[0.0958, -0.0507]^T$	-46.6029	46.6029	
8	10^8	1/8	$[0.1045, 0.9583]^T$	$[1.8967, 2.5251]^T$	-1
	$[0.0958, -0.0507]^T$	$[0.0131, 0.1198]^T$	-47.0771	47.0771	
9	10^9	1/9	$[0.0986, -0.3392]^T$	$[1.9097, 2.6449]^T$	-1
	$[0.0131, 0.1198]^T$	$[0.0110, -0.0377]^T$	-48.0048	48.0048	
10	10^{10}	1/10	$[0.2387, -0.2787]^T$	$[1.9207, 2.6072]^T$	1
	$[0.0110, -0.0377]^T$	$[0.0239, -0.0279]^T$	-47.8419	47.8419	

(continued)

Table 3.11 (continued)

i	α	a	$\boldsymbol{\xi}$	$\mathbf{x}_{(i)}$	sgn
	$\Delta\mathbf{x}_{(i-1)}$	$\Delta\mathbf{x}_{(i)}$	$p(\mathbf{x}_{(i)})$	$f(\mathbf{x}_{(i)})$	$\Delta f(\mathbf{x}_{(i-1)})$
11	10^{11}	1/11	$[0.5130, -0.1722]^T$	$[1.9912, 2.5793]^T$	-1
	$[0.0239, -0.0279]^T$	$[0.0466, -0.0157]^T$	-47.8496	47.8496	
12	10^{12}	1/12	$[-0.0153, 0.3895]^T$	$[1.9912, 2.5637]^T$	-1
	$[0.0466, -0.0157]^T$	$[-0.0013, -0.0325]^T$	-48.1213	48.1213	
13	10^{13}	1/13	$[0.9455, -0.3445]^T$	$[1.9899, 2.5961]^T$	-1
	$[-0.0013, -0.0325]^T$	$[0.0727, -0.0265]^T$	-48.3331	48.3331	
14	10^{14}	1/14	$[0.6756, 0.4781]^T$	$[2.0627, 2.5696]^T$	-1
	$[0.0727, -0.0265]^T$	$[0.0483, 0.0342]^T$	-48.7261	48.7261	
15	10^{15}	1/15	$[0.9083, -0.9362]^T$	$[2.1109, 2.6038]^T$	-1
	$[0.0483, 0.0342]^T$	$[0.0342, 0.0606]^T$	-49.3286	49.3286	

remain relatively far from the actual minimum. Moreover, because the backstep in case of failure returns the solution estimate back *exactly* to the previous position, the solution convergence is very slow. The best data from 20 testing runs follows.

It was computed that

$$\Delta\mathbf{x}_{(1)} = a_{(1)}\boldsymbol{\xi}_{(1)} = \begin{bmatrix} 0.7240 \\ 0.9797 \end{bmatrix} \Rightarrow \mathbf{x}_{(2)} = \begin{bmatrix} 1.7240 \\ 1.9797 \end{bmatrix}$$

with $p\left(\mathbf{x}_{(2)}, 100\right) = -f\left(\mathbf{x}_{(2)}\right) = -41.4768 \Rightarrow \operatorname{sgn}\Delta p\left(\mathbf{x}_{(2)}, 100\right) = -1$, i.e., a successful step. Hence, the next step performs a random move again, giving rise to

$$\Delta\mathbf{x}_{(2)} = 0.5\boldsymbol{\xi}_{(2)} = \begin{bmatrix} 0.0282 \\ 0.7686 \end{bmatrix} \Rightarrow \mathbf{x}_{(3)} = \begin{bmatrix} 1.7384 \\ 2.3640 \end{bmatrix}$$
$$\Rightarrow p\left(\mathbf{x}_{(3)}, 10^3\right) = -f\left(\mathbf{x}_{(3)}\right) = -44.5479 \Rightarrow \operatorname{sgn}\Delta p(\cdot) = -1$$

However, after another random move, it is found that

$$\mathbf{x}_{(4)} = \begin{bmatrix} 1.7971 \\ 2.1339 \end{bmatrix} \Rightarrow p\left(\mathbf{x}_{(4)}, 10^4\right) = -f\left(\mathbf{x}_{(4)}\right) = -43.3471 \Rightarrow \operatorname{sgn}\Delta p(\cdot) = 1$$

That is, this unsatisfactory iteration step implies the return move to $\mathbf{x}_{(3)}$. The data is summarized in Table 3.12.

It is clear that a solution estimate close to the actual extremum will be found; however, at the cost of a high number of iteration steps.

(c) Numerical observations about the adaptive search method with a linear recalculation in this example can be summarized as follows. Again, it is a defensive method yielding the lowest variance of the results of all 20 iteration runs. The method provides the best average cost function value at the minimum point estimate. Conversely, the best (selected) run presented below does not reach values obtained in (a) and (b).

The first iteration step gives

$$\Delta\mathbf{x}_{(1)} = \boldsymbol{\xi}_{(1)} = \begin{bmatrix} -0.8873 \\ -0.6950 \end{bmatrix} \Rightarrow \mathbf{x}_{(2)} = \begin{bmatrix} 0.1127 \\ 0.3050 \end{bmatrix} \Rightarrow p\left(\mathbf{x}_{(2)}, 100\right)$$
$$= -f\left(\mathbf{x}_{(2)}\right) = -5.3446$$

which means that this step fails. Hence, the method dictates to move “behind” $\mathbf{x}_{(1)}$ to the opposite point of $\mathbf{x}_{(2)}$

Table 3.12 The use of the adaptive search method for a cost function with a backstep—Example 3.17

i	α	a	$\boldsymbol{\xi}$	$\mathbf{x}_{(i)}$	sgn
	$\Delta\mathbf{x}_{(i-1)}$	$\Delta\mathbf{x}_{(i)}$	$p(\mathbf{x}_{(i)})$	$f(\mathbf{x}_{(i)})$	$\Delta f(\mathbf{x}_{(i-1)})$
1	10	1	$[0.7240, 0.9797]^T$	$[1, 1]^T$	–
	–	$[0.7240, 0.9797]^T$	– 25	25	
2	100	0.5	$[0.0288, 0.7686]^T$	$[1.7240, 1.9797]^T$	– 1
	$[0.7240, 0.9797]^T$	$[0.0144, 0.3843]^T$	– 41.4768	41.4768	
3	1000	1/3	$[0.1761, -0.6905]^T$	$[1.7384, 2.3640]^T$	– 1
	$[0.0144, 0.3843]^T$	$[0.0587, -0.2302]^T$	– 44.5499	44.5499	
4	10^4	1/4	–	$[1.7971, 2.1339]^T$	1
	$[0.0587, -0.2302]^T$	$[-0.0587, 0.2302]^T$	– 43.3471	43.3471	
5	10^5	1/5	$[0.4974, 0.6512]^T$	$[1.7384, 2.3640]^T$	– 1
	$[-0.0587, 0.2302]^T$	$[0.0995, 0.1302]^T$	– 44.5499	44.5499	
6	10^6	1/6	$[0.5799, -0.3630]^T$	$[1.8379, 2.4943]^T$	– 1
	$[0.0995, 0.1302]^T$	$[0.0967, -0.0605]^T$	– 46.3601	46.3601	
7	10^7	1/7	$[0.0681, -0.8201]^T$	$[1.9345, 2.4338]^T$	– 1
	$[0.0967, -0.0605]^T$	$[0.0097, -0.1172]^T$	– 46.7496	46.7496	
8	10^8	1/8	–	$[1.9443, 2.3166]^T$	1
	$[0.0097, -0.1172]^T$	$[-0.0097, 0.1172]^T$	– 45.9805	45.9805	
9	10^9	1/9	$[0.3573, -0.0096]^T$	$[1.9345, 2.4338]^T$	– 1
	$[-0.0097, 0.1172]^T$	$[0.0397, -0.0011]^T$	– 46.7496	46.7496	
10	10^{10}	1/10	$[-0.6206, -0.0100]^T$	$[1.9742, 2.4327]^T$	– 1
	$[0.0397, -0.0011]^T$	$[-0.0621, -0.0010]^T$	– 47.0668	47.0668	

(continued)

Table 3.12 (continued)

i	α	a	$\boldsymbol{\xi}$	$\mathbf{x}_{(i)}$	sgn
	$\Delta\mathbf{x}_{(i-1)}$	$\Delta\mathbf{x}_{(i)}$	$p(\mathbf{x}_{(i)})$	$f(\mathbf{x}_{(i)})$	$\Delta f(\mathbf{x}_{(i-1)})$
11	10^{11}	1/11	–	$[1.9122, 2.4317]^T$	1
	$[-0.0621, -0.0010]^T$	$[0.0621, 0.0010]^T$	− 46.5491	46.5491	
12	10^{12}	1/12	$[0.7014, 0.1211]^T$	$[1.9742, 2.4327]^T$	− 1
	$[0.0621, 0.0010]^T$	$[0.0585, 0.0101]^T$	− 47.0668	47.0668	
13	10^{13}	1/13	$[0.8592, 0.3933]^T$	$[2.0327, 2.4428]^T$	− 1
	$[0.0585, 0.0101]^T$	$[0.0661, 0.0303]^T$	− 47.6055	47.6055	
14	10^{14}	1/14	$[0.1656, 0.6308]^T$	$[2.0988, 2.4730]^T$	− 1
	$[0.0661, 0.0303]^T$	$[0.0118, 0.0451]^T$	− 48.3312	48.3312	
15	10^{15}	1/15	$[0.7580, 0.9778]^T$	$[2.1106, 2.5181]^T$	− 1
	$[0.0118, 0.0451]^T$	$[0.0505, 0.0652]^T$	− 48.7367	48.7367	

$$\Delta\mathbf{x}_{(2)} = -2\Delta\mathbf{x}_{(1)} = \begin{bmatrix} 1.7746 \\ 1.3900 \end{bmatrix} \Rightarrow \mathbf{x}_{(3)} = \begin{bmatrix} 1.8873 \\ 1.6950 \end{bmatrix}$$
$$\Rightarrow p\left(\mathbf{x}_{(3)}, 1000\right) = -f\left(\mathbf{x}_{(3)}\right) = -40.5401$$

As it is a successful step ($\text{sgn}\Delta p\left(\mathbf{x}_{(3)}, 1000\right) = -1$), the new solution estimate is found randomly in the 4th iteration step. Data of all the iterations is presented in Table 3.13.

(d) The adaptive search method with a prediction is representative of "offensive" (i.e., more aggressive) methods. In case of a successful step, it follows the satisfactory direction; otherwise, the simple random search is applied. Therefore, the numerical solution estimations often leave the constraint set after searching a very good estimate near the actual extremum (that lies on the constraint border). This case is demonstrated in Table 3.14 where a series of linear "jumps" can be observed, while the best solution (out of 20 runs) that eventually satisfies the constraints is summarized in Table 3.15. Again, some steps are commented in detail.

Notice that the solution estimate leaves the constraint set in step $i = 15$.

A "less aggressive" result follows for comparison. The first iterative step gives

$$\Delta\mathbf{x}_{(1)} = \boldsymbol{\xi}_{(1)} = \begin{bmatrix} -0.4448 \\ 0.3050 \end{bmatrix} \Rightarrow \mathbf{x}_{(2)} = \begin{bmatrix} 0.5552 \\ 1.3050 \end{bmatrix},$$
$$p\left(\mathbf{x}_{(2)}, 100\right) = -f\left(\mathbf{x}_{(2)}\right) = -22.2244 \Rightarrow \text{sgn}\Delta p\left(\mathbf{x}_{(2)}, 100\right) = 1$$

As the cost function value has increased, this unsuccessful step is not repeated, and hence, a new random direction is generated in the next step

$$\Delta\mathbf{x}_{(2)} = 0.5\boldsymbol{\xi}_{(2)} = 0.5\begin{bmatrix} 0.8346 \\ 0.0197 \end{bmatrix} = \begin{bmatrix} 0.4173 \\ 0.0098 \end{bmatrix} \Rightarrow \mathbf{x}_{(3)} = \begin{bmatrix} 0.9725 \\ 1.3149 \end{bmatrix},$$
$$p\left(\mathbf{x}_{(3)}, 1000\right) = -f\left(\mathbf{x}_{(3)}\right) = -27.7184 \Rightarrow \text{sgn}\Delta p\left(\mathbf{x}_{(3)}, 1000\right) = -1$$

which implies a better solution estimate. Therefore, the next solution update follows this successful direction again

$$\Delta\mathbf{x}_{(3)} = \Delta\mathbf{x}_{(2)} = \begin{bmatrix} 0.4173 \\ 0.0098 \end{bmatrix} \Rightarrow \mathbf{x}_{(4)} = \begin{bmatrix} 1.3898 \\ 1.3247 \end{bmatrix}$$

The remaining solution data is provided to readers in Table 3.15.

Notice that the minimum estimate has crossed the constraint border in step $i = 9$; however, it has returned back to the admissible region and remained there in the next steps.

Table 3.13 The use of the adaptive search method for a cost function with a linear recalculation—Example 3.17

i	α	a	$\boldsymbol{\xi}$	$\mathbf{x}_{(i)}$	sgn
	$\Delta\mathbf{x}_{(i-1)}$	$\Delta\mathbf{x}_{(i)}$	$p(\mathbf{x}_{(i)})$	$f(\mathbf{x}_{(i)})$	$\Delta f(\mathbf{x}_{(i-1)})$
1	10	1	$[-0.8873, -0.6950]^T$	$[1, 1]^T$	–
	–	$[-0.8873, -0.6950]^T$	-25	25	
2	100	0.5	–	$[0.1127, 0.3050]^T$	1
	$[-0.8873, -0.6950]^T$	$[1.7746, 1.3900]^T$	-5.3446	5.3446	
3	1000	1/3	$[0.6644, 0.2348]^T$	$[1.8873, 1.6950]^T$	-1
	$[1.7746, 1.3900]^T$	$[0.2215, 0.0783]^T$	-40.5401	40.5401	
4	10^4	1/4	$[0.0403, 0.7277]^T$	$[2.1088, 1.7733]^T$	-1
	$[0.2215, 0.0783]^T$	$[0.0101, 0.1819]^T$	-42.9813	42.9813	
5	10^5	1/5	$[-0.8046, 0.8161]^T$	$[2.1189, 1.9552]^T$	-1
	$[0.0101, 0.1819]^T$	$[-0.1609, 0.1632]^T$	-44.5622	44.5622	
6	10^6	1/6	$[-0.7840, 0.0340]^T$	$[1.9579, 2.1184]^T$	-1
	$[-0.1609, 0.1632]^T$	$[-0.1307, 0.0057]^T$	-44.5933	44.5933	
7	10^7	1/7	–	$[1.8273, 2.1241]^T$	1
	$[-0.1307, 0.0057]^T$	$[0.2613, -0.0113]^T$	-43.5358	43.5358	
8	10^8	1/8	$[-0.9908, 0.5334]^T$	$[2.0886, 2.1127]^T$	-1
	$[0.2613, -0.0113]^T$	$[-0.1239, 0.0667]^T$	-45.5824	45.5824	
9	10^9	1/9	–	$[1.9647, 2.1794]^T$	1
	$[-0.1239, 0.0667]^T$	$[0.2477, -0.1333]^T$	-45.1186	45.1186	
10	10^{10}	1/10	$[0.9739, 0.0103]^T$	$[2.2125, 2.0461]^T$	-1
	$[0.2477, -0.1333]^T$	$[0.974, 0.0010]^T$	-45.9759	45.9759	

(continued)

Table 3.13 (continued)

i	α	a	$\boldsymbol{\xi}$	$\mathbf{x}_{(i)}$	sgn
	$\Delta\mathbf{x}_{(i-1)}$	$\Delta\mathbf{x}_{(i)}$	$p(\mathbf{x}_{(i)})$	$f(\mathbf{x}_{(i)})$	$\Delta f(\mathbf{x}_{(i-1)})$
11	10^{11}	1/11	$[-0.4572, -0.7985]^T$	$[2.3098, 2.0471]^T$	-1
	$[0.974, 0.0010]^T$	$[-0.0416, -0.0726]^T$	-46.6614	46.6614	
12	10^{12}	1/12	–	$[2.2683, 1.9745]^T$	1
	$[-0.0416, -0.0726]^T$	$[0.0831, 0.1452]^T$	-45.7978	45.7978	
13	10^{13}	1/13	$[0.5258, -0.8341]^T$	$[2.3514, 2.1197]^T$	-1
	$[0.0831, 0.1452]^T$	$[0.0404, -0.0642]^T$	-47.5075	47.5075	
14	10^{14}	1/14	–	$[2.3919, 2.0555]^T$	1
	$[0.0831, 0.1452]^T$	$[-0.0809, 0.1283]^T$	-47.2689	47.2689	
15	10^{15}	1/15	$[-0.6579, 0.8771]^T$	$[2.3110, 2.1839]^T$	-1
	$[-0.0809, 0.1283]^T$	$[-0.0439, 0.0585]^T$	-47.7314	47.7314	

Note that the sequence of solution estimates does not leave the constraint set in most cases for the methods in (b) and (c). It is mainly due to their defensive manner.

Table 3.14 The use of the adaptive search method for a cost function with a prediction—an aggressive result—Example 3.17

i	α	a	$\boldsymbol{\xi}$	$\mathbf{x}_{(i)}$	sgn
	$\Delta\mathbf{x}_{(i-1)}$	$\Delta\mathbf{x}_{(i)}$	$p(\mathbf{x}_{(i)})$	$f(\mathbf{x}_{(i)})$	$\Delta f(\mathbf{x}_{(i-1)})$
1	10	1	$[0.0890, 0.2129]^T$	$[1, 1]^T$	–
	–	$[0.0890, 0.2129]^T$	– 25	25	
2	100	0.5	–	$[1.0890, 1.2129]^T$	– 1
	$[0.0890, 0.2129]^T$	$[0.0890, 0.2129]^T$	– 28.1362	28.1362	
3	1000	1/3	–	$[1.1781, 1.4258]^T$	– 1
	$[0.0890, 0.2129]^T$	$[0.0890, 0.2129]^T$	– 31.1500	31.1500	
4	10^4	1/4	–	$[1.2671, 1.6387]^T$	– 1
	$[0.0890, 0.2129]^T$	$[0.0890, 0.2129]^T$	– 34.0415	34.0415	
5	10^5	1/5	–	$[1.3562, 1.8515]^T$	– 1
	$[0.0890, 0.2129]^T$	$[0.0890, 0.2129]^T$	– 36.8106	36.8106	
6	10^6	1/6	–	$[1.4452, 2.0644]^T$	– 1
	$[0.0890, 0.2129]^T$	$[0.0890, 0.2129]^T$	– 39.4574	39.4574	
7	10^7	1/7	–	$[1.5343, 2.2773]^T$	– 1
	$[0.0890, 0.2129]^T$	$[0.0890, 0.2129]^T$	– 41.9818	41.9818	
8	10^8	1/8	–	$[1.6233, 2.4902]^T$	– 1
	$[0.0890, 0.2129]^T$	$[0.0890, 0.2129]^T$	– 44.3839	44.3839	
9	10^9	1/9	–	$[1.7124, 2.7031]^T$	– 1
	$[0.0890, 0.2129]^T$	$[0.0890, 0.2129]^T$	– 46.6636	46.6636	
10	10^{10}	1/10	–	$[1.8014, 2.9160]^T$	– 1
	$[0.0890, 0.2129]^T$	$[0.0890, 0.2129]^T$	– 48.8210	48.8210	
11	10^{11}	1/11	–	$[1.8904, 3.1288]^T$	– 1
	$[0.0890, 0.2129]^T$	$[0.0890, 0.2129]^T$	– 50.8560	50.8560	
12	10^{12}	1/12	–	$[1.9795, 3.3417]^T$	– 1
	$[0.0890, 0.2129]^T$	$[0.0890, 0.2129]^T$	– 52.7686	52.7686	
13	10^{13}	1/13	–	$[2.0685, 3.5546]^T$	– 1
	$[0.0890, 0.2129]^T$	$[0.0890, 0.2129]^T$	– 54.5589	54.5589	
14	10^{14}	1/14	–	$[2.1576, 3.7675]^T$	– 1
	$[0.0890, 0.2129]^T$	$[0.0890, 0.2129]^T$	– 56.2268	56.2268	
15	10^{15}	1/15	$[-0.1823, 0.4160]^T$	$[2.2466, 3.9804]^T$	1
	$[0.0890, 0.2129]^T$	$[-0.0122, 0.0277]^T$	$5.1527 \cdot 10^{13}$	57.7724	

(e) In contrast to the preceding method, the adaptive search method with a penalization invokes a rather more aggressive random search direction in case of an unsuccessful step. Therefore, big changes in the solution estimates appear. As a consequence, inconsistent results with significant variances between them are

Table 3.15 The use of the adaptive search method for a cost function with a prediction—a less aggressive result—Example 3.17

i	α	a	$\boldsymbol{\xi}$	$\mathbf{x}_{(i)}$	sgn
	$\Delta\mathbf{x}_{(i-1)}$	$\Delta\mathbf{x}_{(i)}$	$p(\mathbf{x}_{(i)})$	$f(\mathbf{x}_{(i)})$	$\Delta f(\mathbf{x}_{(i-1)})$
1	10	1	$[-0.4448, 0.3050]^T$	$[1, 1]^T$	–
	–	$[-0.4448, 0.3050]^T$	-25	25	
2	100	0.5	$[0.8346, 0.0197]^T$	$[0.5552, 1.3050]^T$	1
	$[-0.4448, 0.3050]^T$	$[0.4173, 0.0098]^T$	-22.2244	22.2244	
3	1000	1/3		$[0.9725, 1.3149]^T$	-1
	$[0.4173, 0.0098]^T$	$[0.4173, 0.0098]^T$	-27.7184	27.7184	
4	10^4	1/4	–	$[1.3898, 1.3247]^T$	-1
	$[0.4173, 0.0098]^T$	$[0.4173, 0.0098]^T$	-32.5157	32.5157	
5	10^5	1/5	–	$[1.8071, 1.3346]^T$	-1
	$[0.4173, 0.0098]^T$	$[0.4173, 0.0098]^T$	-36.6162	36.6162	
6	10^6	1/6	–	$[2.2244, 1.3444]^T$	-1
	$[0.4173, 0.0098]^T$	$[0.4173, 0.0098]^T$	-40.0200	40.0200	
7	10^7	1/7	–	$[2.6417, 1.3542]^T$	-1
	$[0.4173, 0.0098]^T$	$[0.4173, 0.0098]^T$	-42.7270	42.7270	
8	10^8	1/8	–	$[3.0590, 1.3641]^T$	-1
	$[0.4173, 0.0098]^T$	$[0.4173, 0.0098]^T$	-44.7373	44.7373	
9	10^9	1/9	$[-0.7889, 0.1877]^T$	$[3.4763, 1.3739]^T$	1
	$[0.4173, 0.0098]^T$	$[-0.0877, 0.0209]^T$	$1.0485 \cdot 10^7$	46.0509	
10	10^{10}	1/10	–	$[3.3887, 1.3948]^T$	-1
	$[-0.0877, 0.0209]^T$	$[-0.0877, 0.0209]^T$	-46.0444	46.0444	
11	10^{11}	1/11	$[-0.9987, -0.4328]^T$	$[3.3010, 1.4156]^T$	1

(continued)

Table 3.15 (continued)

i	α	a	$\boldsymbol{\xi}$	$\mathbf{x}_{(i)}$	sgn
	$\Delta\mathbf{x}_{(i-1)}$	$\Delta\mathbf{x}_{(i)}$	$p(\mathbf{x}_{(i)})$	$f(\mathbf{x}_{(i)})$	$\Delta f(\mathbf{x}_{(i-1)})$
	$[-0.0877, 0.0209]^T$	$[-0.0908, -0.0393]^T$	-46.0063	46.0063	
12	10^{12}	1/12	$[0.1016, 0.7418]^T$	$[3.2102, 1.3763]^T$	1
	$[-0.0908, -0.0393]^T$	$[0.0085, 0.0618]^T$	-45.3736	45.3736	
13	10^{13}	1/13		$[3.2187, 1.4381]^T$	-1
	$[0.0085, 0.0618]^T$	$[0.0085, 0.0618]^T$	-45.9681	45.9681	
14	10^{14}	1/14	–	$[3.2272, 1.4999]^T$	-1
	$[0.0085, 0.0618]^T$	$[0.0085, 0.0618]^T$	-46.5546	46.5546	
15	10^{15}	1/15	–	$[3.3256, 1.5617]^T$	-1
	$[0.0085, 0.0618]^T$	$[0.0085, 0.0618]^T$	-47.1331	47.1331	

obtained (out of 20 computation runs), similar to those obtained by the basic search method. The best obtained data follows.

It is computed that

$$\Delta \mathbf{x}_{(1)} = \boldsymbol{\xi}_{(1)} = \begin{bmatrix} 0.0472 \\ -0.4024 \end{bmatrix} \Rightarrow \mathbf{x}_{(2)} = \begin{bmatrix} 1.0472 \\ 0.5976 \end{bmatrix},$$
$$p\left(\mathbf{x}_{(2)}, 100\right) = -f\left(\mathbf{x}_{(2)}\right) = -21.3762 \Rightarrow \operatorname{sgn}\Delta p\left(\mathbf{x}_{(2)}, 100\right) = 1$$

The first step has failed; therefore, the next move is computed via random $\boldsymbol{\xi}_{(2)}$ as

$$\boldsymbol{\xi}_{(2)} = \begin{bmatrix} 0.4079 \\ -0.2368 \end{bmatrix} \Rightarrow \Delta \mathbf{x}_{(2)} = 0.5\boldsymbol{\xi}_{(2)} \odot \operatorname{sgn}\Delta \mathbf{x}_{(1)}$$
$$= \begin{bmatrix} 0.4079 \\ -0.2368 \end{bmatrix} \odot \begin{bmatrix} 1 \\ -1 \end{bmatrix} = \begin{bmatrix} 0.2040 \\ 0.1184 \end{bmatrix}$$

Since $p\left(\mathbf{x}_{(3)}, 1000\right) = -f\left(\mathbf{x}_{(3)}\right) = -24.9672 \Rightarrow \operatorname{sgn}\Delta p\left(\mathbf{x}_{(3)}, 1000\right) = -1$, the next iterative steps follow this successful direction. The remaining data are summarized in Table 3.16.

Apparently, the best results obtained for the random search methods (within the number of 20 run at 15 (14) iteration steps for each of the methods) in this example do not reach the result obtained for the Nelder–Mead method in Example 3.15. The particular data is eventually summarized in Table 3.17 for visualization. Recall that the actual (analytic) minimum is at $\mathbf{x}^* = [8/3, 10/3]^T = \left[2.\overline{6}, 3.\overline{3}\right]^T$ with the corresponding objective function value $f(\mathbf{x}^*) = 57.\overline{3}$.

Random search methods suffer from a poor and relatively obtuse search within a solution space. It might be enhanced by a metaoptimization setting of control parameters a and ξ, see (3.96).

Table 3.16 The use of the adaptive search method for a cost function with a penalization—Example 3.17

i	α / $\Delta\mathbf{x}_{(i-1)}$	a / $\Delta\mathbf{x}_{(i)}$	$\boldsymbol{\xi}$ / $p(\mathbf{x}_{(i)})$	$\mathbf{x}_{(i)}$ / $f(\mathbf{x}_{(i)})$	sgn $\Delta f(\mathbf{x}_{(i-1)})$
1	10	1	$[0.0472, -0.4024]^T$	$[1, 1]^T$	–
	–	$[0.0472, -0.4024]^T$	-25	25	
2	100	0.5	$[0.4079, -0.2368]^T$	$[1.0472, 0.5976]^T$	1
	$[0.0472, -0.4024]^T$	$[0.2040, 0.1184]^T$	-21.3762	21.3762	
3	1000	1/3	–	$[1.2512, 0.7160]^T$	-1
	$[0.2040, 0.1184]^T$	$[0.2040, 0.1184]^T$	-24.9672	24.9672	
4	10^4	1/4	–	$[1.4551, 0.8344]^T$	-1
	$[0.2040, 0.1184]^T$	$[0.2040, 0.1184]^T$	-28.3639	28.3639	
5	10^5	1/5	–	$[1.6591, 0.9528]^T$	-1
	$[0.2040, 0.1184]^T$	$[0.2040, 0.1184]^T$	-31.5660	31.5660	
6	10^6	1/6	–	$[1.8631, 1.0712]^T$	-1
	$[0.2040, 0.1184]^T$	$[0.2040, 0.1184]^T$	-34.5738	34.5738	
7	10^7	1/7	–	$[2.0670, 1.1896]^T$	-1
	$[0.2040, 0.1184]^T$	$[0.2040, 0.1184]^T$	-37.3871	37.3871	
8	10^8	1/8	–	$[2.2710, 1.3080]^T$	-1
	$[0.2040, 0.1184]^T$	$[0.2040, 0.1184]^T$	-40.0059	40.0059	
9	10^9	1/9	–	$[2.4750, 1.4264]^T$	-1
	$[0.2040, 0.1184]^T$	$[0.2040, 0.1184]^T$	-42.4303	42.4303	
10	10^{10}	1/10	–	$[2.6789, 1.5447]^T$	-1
	$[0.2040, 0.1184]^T$	$[0.2040, 0.1184]^T$	-44.6603	44.6603	
11	10^{11}	1/11	–	$[2.8829, 1.6631]^T$	-1

(continued)

Table 3.16 (continued)

i	α	a	$\boldsymbol{\xi}$	$\mathbf{x}_{(i)}$	sgn
	$\Delta\mathbf{x}_{(i-1)}$	$\Delta\mathbf{x}_{(i)}$	$p(\mathbf{x}_{(i)})$	$f(\mathbf{x}_{(i)})$	$\Delta f(\mathbf{x}_{(i-1)})$
	$[0.2040, 0.1184]^T$	$[0.2040, 0.1184]^T$	− 46.0063	46.0063	
12	10^{12}	1/12	–	$[3.0869, 1.7815]^T$	− 1
	$[0.2040, 0.1184]^T$	$[0.2040, 0.1184]^T$	− 48.5368	48.5368	
13	10^{13}	1/13		$[3.2908, 1.8999]^T$	− 1
	$[0.2040, 0.1184]^T$	$[0.2040, 0.1184]^T$	− 50.1835	50.1835	
14	10^{14}	1/14	–	$[3.4948, 2.0183]^T$	− 1
	$[0.2040, 0.1184]^T$	$[0.2040, 0.1184]^T$	− 51.6356	51.6356	
15	10^{15}	1/15	–	$[3.6988, 2.1367]^T$	− 1
	$[0.2040, 0.1184]^T$	$[0.2040, 0.1184]^T$	− 52.8934	52.8934	

Table 3.17 A comparison of the Nelder–Mead method and the search methods best solution estimates for a cost function with a penalization—Examples 3.15 and 3.17

Method	$\mathbf{x}^*$	$p(\mathbf{x}^*, \cdot)$	$f(\mathbf{x}^*)$
Nelder–Mead	$[2.434, 3.520]^T$	– 56.945	56.945
Simple search	$[2.111, 2.604]^T$	– 49.329	49.329
Adaptive search with a backstep	$[2.111, 2.518]^T$	– 48.737	48.737
Adaptive search with a linear recalculation	$[2.311, 2.184]^T$	– 47.731	47.731
Adaptive search with a prediction	$[2.158, 3.768]^T([3.326, 1.562]^T)$	– 56.227 (– 47.1331)	56.227 (47.1331)
Adaptive search with a penalization	$[3.699, 2.137]^T$	– 52.8934	52.8934

References

1. Avriel, M.: Nonlinear Programming: Analysis and Methods. Dover Publications, New York (2003)
2. Baba, N.: Convergence of a Random Optimization Method for Constrained Optimization Problems. J. Optim. Theory Appl. **33**(4), 451–461 (1981). https://doi.org/10.1007/bf00935752
3. Bertsekas, D.P.: Constrained Optimization and Lagrange Multiplier Methods. Athena Scientific, Nashua (1996)
4. Blum, C., Roli, A.: Metaheuristics in combinatorial optimization: overview and conceptual comparison. ACM Comput. Surv. **35**(3), 268–308 (2003). https://doi.org/10.1145/937503.937505
5. Bonnans, J.F., Gilbert, J.C., Lemaréchal, C., Sagastizábal, C.A.: Numerical Optimization: Theoretical and Practical Aspects, 2nd edn. Springer, Berlin (2006)
6. Chong, E.K.P., Lu, W.S., Zak, S.H.: An Introduction to Optimization, 5th edn. Wiley, New York (2023)
7. Fischetti, M.: Introduction to Mathematical Optimization. Independently Published (2019)
8. Fletcher, R.: Practical Methods of Optimization, 2nd edn. Wiley, New York (1987)
9. French, M.: Fundamentals of Optimization: Methods, Minimum Principles, and Applications for Making Things Better. Springer, Cham (2018)
10. Gill, P.E., Murray, W., Wright, M.H.: Practical Optimization. Academic Press, London (1981)
11. Guenin, B., Könemann, J., Tunçel, L.: A Gentle Introduction to Optimization. Cambridge University Press, Cambridge (2014)
12. Hillier, F.S., Lieberman, G.J.: Introduction to Operations Research, 7th edn. McGraw-Hill, New York (2002)
13. Kowalik, J., Osborne, M.R.: Methods for Unconstrained Optimization Problems. American Elsevier, New York (1968)
14. Lange, K.: Optimization, 2nd edn. Springer, New York (2013)
15. Meisner, G.B., Araujo, R.: The Golden Ratio: The Divine Beauty of Mathematics. Race Point Publishing, London (2018)
16. Nelder, J.A., Mead, R.: A Simplex Method for Function Minimization. Comput. J. **7**(4), 308–313 (1965). https://doi.org/10.1093/comjnl/7.4.308
17. Nocedal, J., Wright, S.J.: Numerical Optimization, 2nd edn. Springer, New York (2006)
18. Rardin, R.L.: Optimization in Operations Research, 2nd edn. Pearson, London (2016)
19. Snyman, J.A., Wilke, D.N.: Practical Mathematical Optimization: Basic Optimization Theory and Gradient-Based Algorithms 2. Springer, Cham (2019)

20. Solis, F.J., Wets, R.J.B.: Minimization by random search techniques. Math. Oper. Res. **6**(1), 19–30 (1981). https://doi.org/10.1287/moor.6.1.19
21. Spendley, W., Hext, G.R., Himsworth, F.R.: Sequential application of simplex designs in optimisation and evolutionary operation. Technometrics **4**, 441–461 (1962). https://doi.org/10.1080/00401706.1962.10490033
22. Tabak, D., Kuo, B.C.: Optimal Control by Mathematical Programming. Prentice-Hall, New York (1971)
23. Walsh, G.R.: Methods of Optimization. Wiley, New York (1975)
24. Winston, W.L.: Operations Research: Applications and Algorithms. Brooks/Cole—Thomson Learning, Belmond (2004)

Chapter 4
Linear Programming

Linear programming (LP) represents the best-developed family of methods solving specific constrained optimization problems. These tasks consider a linear objective multivariable function with constraints composed of linear inequalities. They represent models of various real-life and economic problems. For instance, a possible task can be determining the production level to maximize the revenues when constraining resources, storage facilities, employees, working hours, etc. The goal can also be to minimize the costs to reach the desired guaranteed minimal production level. Another model can represent the so-called mixing problem: making the cheapest mixture containing minimal portions of necessary substances. Using LP, one can also solve problems of material parting, i.e., producing the required number of pieces of particular products with a minimum of the given material resources. Last but not least, some transportation (traffic) problems can be solved using LP techniques to distribute products to the shops at a minimum cost.

In this chapter, a concise historical overview is given to readers, the basic LP optimization task is introduced, and some preliminary notions and statements are presented first. Then, a graphical solution for a two-dimensional case and the simplex-table method—a classical LP technique—are provided in detail for the standard (fundamental) problem. The simplex-table method can be extended to several types of constraint inequalities, giving rise to the mixed-constraint problem of LP. The so-called dual model of LP that enables the transformation of a specific minimization problem into a maximization one is also presented. As apparent from the above-given examples of real-world problems, LP often needs to solve tasks requiring an integer-valued result (e.g., the number of products). The curious readers are referred to the literature [1–17] if they are interested in learning more details about LP and related problems and techniques.

L. Pekař, *Optimization: An Introduction*, Studies in Systems, Decision and Control 239,
https://doi.org/10.1007/978-3-031-86326-4_4

4.1 History, Problem Formulation and Preliminaries

4.1.1 LP History Overview

LP is a relatively young branch of mathematics, the leading development of which began only during World War II. However, the first predecessor dates back to the eighteenth century. The French economist François Quesnay (1694–1774) developed an "economic table" to study the mutual relationship between landlords, farmers, and artisans in 1758. The table enabled us to find non-negative solutions for linear equalities and nonequalities. In the 1820s, a famous Frenchman, Jean Baptiste Joseph Fourier (1768–1830), published an algorithm solving a set of linear equations. Later, at the turn of the nineteenth and twentieth centuries, the Hungarian mathematician Gyula Farkas de Kisbarnak (1847–1930) presented a theory for solving a finite system of linear inequalities. The transportation problem was first formulated by the French mathematician Gaspard Monge (1746–1818) at the end of the eighteenth century. Its solution was proposed by, e.g., A. N. Tolstoi and Frank Lauren Hitchcock (1875–1957) in the 1930s and 1940s. In the 1930s, the matching problem (that seeks the most efficient pairing of two groups of the same size) was also solved, resulting in the so-called Hungarian method. Leonid Vitalyevich Kantorovich (1912–1986) published a book entitled Mathematical Methods in the Organization and Planning of Production in 1939, representing an essential contribution to the LP theory [14]. In 1975, with the economist Tjalling Charles Koopmans (1910–1985), Kantorovich received the Nobel Prize in Economics for his contribution to the development of LP and its economic applications. During World War II, Jerome Cornfield (1912–1979) formulated and approximately solved the diet problem. The primary motivation was finding the cheapest but nutritionally sufficient menu for the soldiers. The most famous LP solution technique, the simple-table method, was George Bernard Dantzig (1914–2005) in 1947; [2] however, economists Tjalling Charles Koopmans and Wassily Wassilyevich Leontief (1906–1999) also participated in its development. For the sake of this chapter, Ralph Edward Gomory (1929) can be named, who proposed the cutting-plane method for solving IPL and mixed IPL problems in the 1950s [17].

4.1.2 Primary LP Problem Formulation

The basic (**primary**) **LP problem** is formulated via the following linear fitness function with the inequality-constraint set

$$\max f(\mathbf{x}) = c_1x_1 + c_2x_2 + \cdots + c_nx_n = \mathbf{c}^T\mathbf{x} \in \mathbb{R} \tag{4.1}$$

s.t.:

$$\left.\begin{aligned} &a_{11}x_1 + a_{12}x_2 + \cdots + a_{1n}x_n \le b_1 \\ &a_{21}x_1 + a_{22}x_2 + \cdots + a_{2n}x_n \le b_2 \\ &\cdots \\ &a_{m1}x_1 + a_{m2}x_2 + \cdots + a_{mn}x_n \le b_m \end{aligned}\right\} \Leftrightarrow \mathbf{Ax} \le \mathbf{b};\ \mathbf{A} \in \mathbb{R}^{m\times n},\ \mathbf{b} \in \mathbb{R}^m,\ \mathbf{x} > \mathbf{0}_n \in \mathbb{R}^n \tag{4.2}$$

where $\mathbf{x}$ is a vector of **decision variables**, $\mathbf{b}$, $\mathbf{c}$ are **coefficient** vectors.

Note that the analytic solution of problem (4.1)–(4.2) can be done using the Karush–Kuhn–Tucker theorem, see Theorem 2.4. However, the LP techniques utilize iterative procedures that yield the exact solution as well.

The primary problem maximizes the objective function, while all the constraints are represented by linear inequalities of the "less or equal" type. Nevertheless, there can exist also other problems extending the basic one, i.e.,

$$\begin{aligned} &\min f(\mathbf{x}),\ \mathbf{Ax} \le \mathbf{b} \\ &\max f(\mathbf{x}),\ \mathbf{Ax} \ge \mathbf{b} \\ &\min f(\mathbf{x}),\ \mathbf{Ax} \ge \mathbf{b} \\ &\mathbf{x} \ge \mathbf{0}_n \end{aligned} \tag{4.3}$$

Tasks (4.3) give rise to the mixed-constraint or "combined" LP problems.

4.1.3 Basic Notions and Statements

Now, let the essential notions and facts related to (4.1)–(4.3) be introduced theoretically and also via demonstrative examples.

Consider a set of inequalities (4.2), which is adjusted to the **canonical equation form** by introducing **slack variables** $x_{n+1}, x_{n+2}, \ldots, x_{n+m} \ge 0$ that express the "slackness" of the left-hand sides of a particular inequality

$$\begin{aligned} a_{11}x_1 + a_{12}x_2 + \cdots + a_{1n}x_n + x_{n+1} &= b_1 \\ a_{21}x_1 + a_{22}x_2 + \cdots + a_{2n}x_n + x_{n+2} &= b_2 \\ &\cdots \\ a_{m1}x_1 + a_{m2}x_2 + \cdots + a_{mn}x_n + x_{n+m} &= b_m \end{aligned} \tag{4.4}$$

i.e., in the matrix form

$$\begin{bmatrix} a_{11} & \cdots & a_{1n} & 1 & 0 & \cdots & 0 \\ \vdots & \ddots & \vdots & \vdots & \vdots & \vdots & \vdots \\ a_{m1} & \cdots & a_{11} & 0 & 0 & \cdots & 1 \end{bmatrix} \begin{bmatrix} x_1 \\ \vdots \\ x_n \\ x_{n+1} \\ \vdots \\ x_{n+m} \end{bmatrix} = \begin{bmatrix} b_1 \\ b_2 \\ \vdots \\ b_m \end{bmatrix}$$

$$\tilde{\mathbf{A}}\tilde{\mathbf{x}} = \mathbf{b} \tag{4.5}$$

If any row of (4.5) includes a negative b_i, the particular row is multiplied by -1.

Hence, (4.4) or (4.5) represents a set of m linear algebraic equations with $m + n$ unknown variables. A solution of this system exits if

$$\operatorname{rank} \tilde{\mathbf{A}} = \operatorname{rank}\left[\, \tilde{\mathbf{A}} \; \mathbf{b} \,\right] \tag{4.6}$$

and the number of these solutions is infinite. Assume that $\operatorname{rank} \tilde{\mathbf{A}} = m$ (i.e., the equation system has full row rank). Note that if $\operatorname{rank} \tilde{\mathbf{A}} < m$, the set can be reduced, and the so-called *degenerate solution* is obtained.

Now, if the number of *n non-basic variables* are selected, the number of m dependent *basic variable* values are obtained. Let all the non-basic variables be set to zero, then the number

$$\binom{n+m}{m} \tag{4.7}$$

of different **basic solutions** can be obtained. That is, the number of possibilities for selecting m columns in $\tilde{\mathbf{A}}$ (i.e., the basic variables) from all $m + n$ columns (i.e., all variables) is given by (4.7).

Definition 4.1 A **basic admissible solution** of (4.4)–(4.5) is a basic solution satisfying $\mathbf{x} \geq \mathbf{0}_n$.

Example 4.1 *Task*: Let the given set of algebraic inequalities be

$$x_1 + 2x_2 \leq 6$$
$$2x_1 + x_2 \leq 5$$

Find all basic solutions and decide which of them are admissible.

Solution: The right-hand side vector $\mathbf{b} = [6, 5]^T$ has all its entries non-negative. Hence, the canonical form reads

$$x_1 + 2x_2 + x_3 = 6$$
$$2x_1 + x_2 + x_4 = 5$$

where x_3, x_4 are slack variables. It can easily be verified that

$$\begin{aligned} \operatorname{rank}\tilde{\mathbf{A}} &= \operatorname{rank}\begin{bmatrix} 1\,2\,1\,0 \\ 2\,1\,0\,1 \end{bmatrix} = \operatorname{rank}\begin{bmatrix} 1\,2\,1\,0\,6 \\ 2\,1\,0\,1\,5 \end{bmatrix} \\ &= \operatorname{rank}\begin{bmatrix} \tilde{\mathbf{A}}\; \mathbf{b} \end{bmatrix} = 2 \end{aligned}$$

hence, the equation system has infinitely many solutions, see (4.6).

According to (4.7), it is possible to find six basic solution vectors by setting each pair of non-basic variables to zero. The procedure can be represented by standard matrix row operations on $\begin{bmatrix} \tilde{\mathbf{A}}\; \mathbf{b} \end{bmatrix}$ where the basic variables are given by *unit submatrix* of $\tilde{\mathbf{A}}$. Then, the corresponding solution values are determined by the current $\mathbf{b}$.

$$\begin{aligned} &\left[\begin{array}{cccc|c} 1&2&1&0&6 \\ 2&1&0&1&5 \end{array}\right] \sim \left[\begin{array}{cccc|c} 1&2&1&0&6 \\ 0&-3&-2&1&-7 \end{array}\right] \\ &\sim \left[\begin{array}{cccc|c} 1&0&-1/3&2/3&4/3 \\ 0&1&2/3&-1/3&7/3 \end{array}\right] \\ &\sim \left[\begin{array}{cccc|c} 1&1/2&0&1/2&5/2 \\ 0&3/2&1&-1/2&7/2 \end{array}\right] \sim \left[\begin{array}{cccc|c} 2&1&0&1&5 \\ -3&0&1&-2&-4 \end{array}\right] \\ &\sim \left[\begin{array}{cccc|c} 1/2&1&1/2&0&3 \\ 3/2&0&-1/2&1&2 \end{array}\right] \end{aligned}$$

The unit submatrices give rise to the following basic solutions

$$\mathbf{x}^1 = \begin{bmatrix} 0 \\ 0 \\ 6 \\ 5 \end{bmatrix}, \mathbf{x}^2 = \begin{bmatrix} 6 \\ 0 \\ 0 \\ -7 \end{bmatrix}, \mathbf{x}^3 = \begin{bmatrix} 4/3 \\ 7/3 \\ 0 \\ 0 \end{bmatrix},$$

$$\mathbf{x}^4 = \begin{bmatrix} 5/2 \\ 0 \\ 7/2 \\ 0 \end{bmatrix}, \mathbf{x}^5 = \begin{bmatrix} 0 \\ 5 \\ -4 \\ 0 \end{bmatrix}, \mathbf{x}^6 = \begin{bmatrix} 0 \\ 3 \\ 0 \\ 2 \end{bmatrix}$$

The admissible basic solutions are $\mathbf{x}^1$, $\mathbf{x}^3$, $\mathbf{x}^4$, and $\mathbf{x}^6$, while $\mathbf{x}^2$ and $\mathbf{x}^5$ are non-admissible ones as they include negative entries.

The basic solutions have a nice *geometrical interpretation*. Whereas all basic solutions are intersections of each equality constraint pair given by the canonical form, the *admissible subset agrees with the vertices of the constraint polyhedron*, see Fig. 4.1.

The **optimal solution** of an LP problem is naturally the admissible solution that extremizes the linear objective function $f(\mathbf{x})$. It is worth noting that the inequality

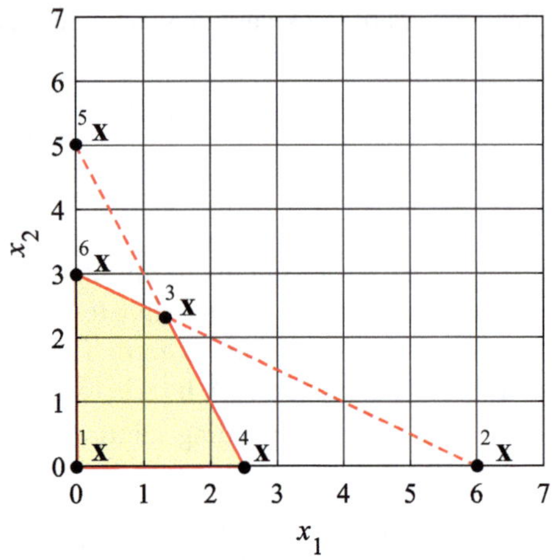

Fig. 4.1 Basic solutions—Example 4.1

constraint system (or the complete LP problem) can have no solution, infinitely many (optimal) solutions, or an (optimal) solution at infinity.

Figure 4.2 elucidates that the constraints set

$$\begin{aligned} x_1 + 2x_2 &\leq 2 \\ 5x_1 + 3x_2 &\geq 15 \end{aligned} \tag{4.8}$$

has no intersection; hence, a solution cannot exist.

Now, consider the LP problem

$$\min f(x_1, x_2) = x_1 + 2.5x_2$$

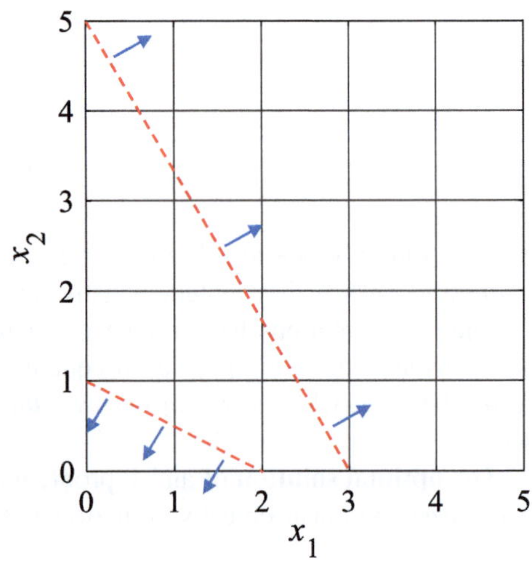

Fig. 4.2 A solution of (4.8) does not exist

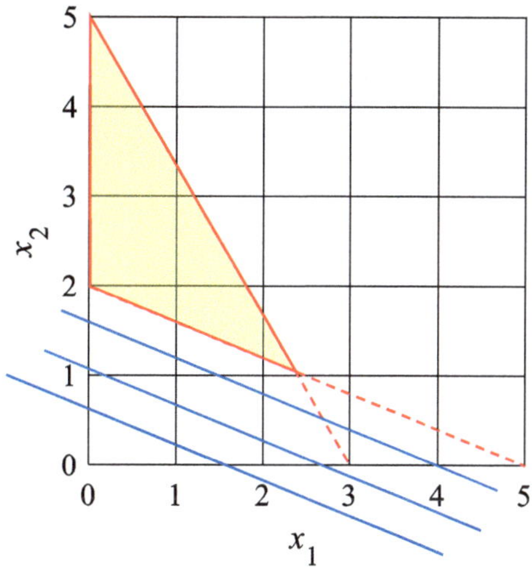

Fig. 4.3 Infinitely many optimal solutions of (4.9)

$$\begin{aligned} \text{s.t. } 2x_1 + 5x_2 &\geq 10 \\ 5x_1 + 3x_2 &\leq 15 \end{aligned} \tag{4.9}$$

The problem is depicted in Fig. 4.3, where the blue lines display contour lines of $f(\mathbf{x})$, i.e., the points at which $f(\mathbf{x}) = \text{const.}$. The line $2x_1 + 5x_2 = 10$, parallel to the blue lines, obviously represents all the infinitely many optimal solutions.

A case of the unbounded optimal solution can be expressed, e.g., by the problem

$$\begin{aligned} \max f(x_1, x_2) &= x_1 + x_2 \\ 2x_1 + 5x_2 &\geq 10 \end{aligned} \tag{4.10}$$

see also Fig. 4.4.

The following crucial theorem (also called the fundamental LP theorem) holds for the optimal solution of an LP problem.

Theorem 4.1 *If* $\mathbf{x}^*$ *is the optimal solution of an LP problem, it is the basic admissible solution ; i.e., it agrees with a vertex of the constraint polyhedron.*

It follows from Theorem 4.1, that it is enough to check objective function values at all the constraint polyhedron vertices. Such a task is simple for a low-dimensional problem with a small set of constraints. However, it is inefficient for a general LP problem. Moreover, many existing methods offer more than just finding an extremum.

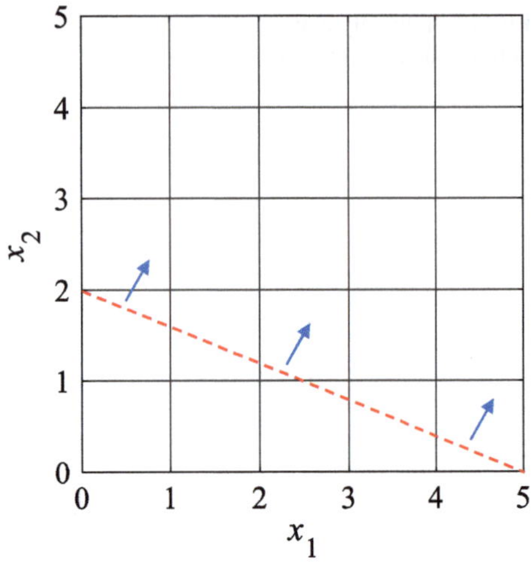

Fig. 4.4 Unbounded optimal solution of (4.10)

4.2 LP Problem Solution Techniques

4.2.1 Graphical Solution

The first solution method is reasonable to be applied to a two-dimensional problem. It is based on the fact that the linear objective function $f(\mathbf{x})$ constitutes a hyperplane in the $\mathbb{R}^{n+1}$ space. In the $\mathbb{R}^2$ space, the function can geometrically be imagined as a plane that forms an angle to axes x_1, x_2. This inclination (i.e., gradient) is represented by vector $\mathbf{c} = [c_1, c_2]^T$. Such a geometrical imagination gives rise to the intuitive validation of Theorem 4.1. It is clear that an extremum lying at a polyhedron vertex must be searched in the gradient direction since the function value the functional value monotonically grows (or decreases) the most.

Hence, if a line tangential to the gradient is made in the plane (see, e.g., the blue lines in Fig. 4.3) and this line is shifted in the gradient direction (i.e., a bunch of mutually parallel lines is drawn), then the farthermost polyhedron vertex crossed by the line constitutes the maximum point.

Example 4.2 *Task*: Solve the following LP problem

$$\begin{aligned} &\max f(x_1, x_2) = x_1 + 2x_2 \\ &\text{s. t. } 3x_1 + 4x_2 \leq 12 \\ &\qquad 5x_1 + 3x_2 \leq 15 \end{aligned}$$

graphically.

Solution: The blue lines in Fig. 4.5 indicate $f(x_1, x_2) = \text{const.}$ (e.g., the bottom-most one agrees with $f(x_1, x_2) = 0$). Hence, the upper-most line crossing a constraint

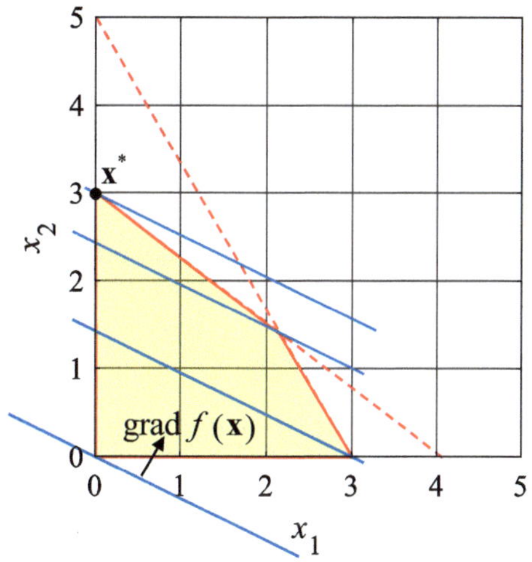

Fig. 4.5 The graphical solution of an LP problem—Example 4.2

polyhedron vertex implies the objective function maximum at $\mathbf{x}^* = [0, 3]^T$ with the function value of $f(\mathbf{x}^*) = 1 \cdot 0 + 2 \cdot 3 = 6$.

4.2.2 Simplex-Table Method for the Primary LP Problem

The simplex-table method represents the essential technique for solving the LP problems. It enables not only the determination of the objective function extremum point and its value but also the analysis of the solution regarding its sensitivity to problem model coefficients or the determination of problem solvability.

Below, the computational procedure for the primary problem (4.1)–(4.2) is presented (see Algorithm 4.1). Note that techniques for tackling problems (4.3) are introduced in Sects. 4.2.3 and 4.2.4.

Algorithm 4.1 *(The primary simplex-table method)*

1. Consider the primary problem (4.1)–(4.2), and transform it in the canonical equation form (4.4)–(4.5) (with a non-negative right-hand-side vector **b**). Hence, each inequality adds a slack variable.
2. The canonical form of the constraints, along with the row $-\mathbf{c}^T$ expressing the objective function, then forms the following initial **simplex table** (Table 4.1).
3. If all coefficients in the last row (except for the rightmost column) are non-negative, the solution is *optimal*, and go to step 7. Else, go to step 4.
4. Find the **pivot column**, the index of which satisfies

$$j^p := \arg\min_j \{\text{numbers in the last row}\} \tag{4.11}$$

Table 4.1 The initial simplex table for the primary problem

x_1	x_2	...	x_n	x_{n+1}	x_{n+2}	...	x_{n+m}	b_i
a_{11}	a_{12}	...	a_{1n}	1	0	...	0	b_1
a_{21}	a_{22}	...	a_{2n}	0	1	...	0	b_2
$\vdots$	$\vdots$	$\vdots$	$\vdots$	$\vdots$	$\vdots$	$\ddots$	$\vdots$	$\vdots$
a_{m1}	a_{m2}	...	a_{mn}	0	0	...	1	b_m
$-c_1$	$-c_2$	...	$-c_n$	0	0	...	0	0

5. Find the **pivot row** index using the **ratio test** with pivot column as

$$i^p := \arg\min_i \left\{ \beta_i = \frac{b_i}{a_{ij^p}} \geq 0 \right\},\ i = 1, 2, \ldots m \tag{4.12}$$

   If all β_i are negative or infinite, the LP problem is *unbounded* (i.e., increasing the value of the entering variable given by j^p leads to an infinite objection function value), and terminate the algorithm. Otherwise, go to step 6.
6. The intersection of the pivot column and the pivot row determines the **pivot element**. This element at position $[i^p, j^p]$ is *eliminated* by the standard row operations (i.e., the pivot row can be multiplied by a real number and added to or subtracted from any other row). The goal of the elimination is to obtain value one at the pivot element position $[i^p, j^p]$ and zeros at all positions $[i, j^p]$, $i = 1, 2, \ldots, n+1 \neq i^p$ (i.e., including the last row but excluding the pivot row). Then, go to step 3.
7. *Optimal results* are deduced. The right-bottom element of the simplex table (i.e., in the last column and row) equals the optimal *objective function value* f^*. The *basic* variables are determined by unit column vectors, and their optimal values are given by the number in the rightmost column and the row where the particular basic variable has one in the unit vector. All other variables are *non-basic*, and their optimal values are zero.

Details to Algorithm 4.1 follow.

- In each step of the table updating, a *unit submatrix* must always exist.
- The current results (albeit non-optimal) can be deduced for each particular form of the simplex table. For instance, the initial form from step 1 of the algorithm (see Table 4.1) agrees with $f = 0, x_1 = x_2 = \ldots = x_n = 0, x_{n+1} = b_1, x_{n+2} = b_2$, ..., $x_{n+m} = b_m$; i.e., all decision variables are zero, and the slack variables are given by the constraints.
- Regarding the elimination operation from step 6 of the algorithm, the standard row operations can be done as follows:
  - The pivot row is multiplied by $1/a_{i^p j^p}$, which yields one at position $[i^p, j^p]$.

- The pivot row is then successively multiplied by $-a_{ij^p}$ for each $i = 1, 2, \dots, n+1 \neq i^p$, and this adjusted row is added to the particular ith row of the table. This operation yields zeros in the pivot column.
- Beware that multiplying a non-pivot row and adding it to another row does not pose a standard row operation!

- If $b_i = 0$ and $a_{ij^p} > 0$ in ratio test (4.12), the corresponding pivot element exists (as a degenerative case). A combination of $b_i = 0$ with $a_{ij^p} < 0$ is unacceptable.
- The optimal form of the simplex table as per step 7 of Algorithm 4.1 provides many useful information:

  - If a basic variable (either a decision or slack one) equals zero, the so-called *degenerated solution* is obtained. It geometrically means that the polyhedron vertex determining the extreme is composed of an intersection of more than two constraints.
  - If for a non-basic variable, there is a zero in the last row of the simplex table, this variable can become basic without changing the objective function value. Hence, *multiple* optimal *solutions* exist.
  - **Shadow prices**: The coefficient vector **b** is often economically interpreted as the limit on material resources, budget, employees, etc. The simplex table says whether these resources are exhausted or not. If so, it is possible to make a *sensitivity analysis* to find out the marginal change in the objective function if particular b_i is changed a bit. This concept is called a *shadow* (or *dual*) *price*. Namely, if a slack (and non-basic) variable equals zero, no "margin" to the constraint exists. The number in the last row for the particular slack variable then agrees with the shadow price. Naturally, if the slack variable is basic (and hence non-zero in general), the resource is not exhausted, and an increment of b_i is useless, which is indicated by the zero in the last row.
  - **Reduced prices**: Numbers in the last row under the decision variables also have special meaning. Namely, a particular non-zero number for a non-basic decision variable expresses how much the coefficient c_j (for some $j = 1, 2, \dots, n$) would have to increase to make the respective variable basic. For instance, if c_j express prices of various product units, it is possible to identify the "minimum" price at which it would be worthwhile to sell the product.

A possible Matlab code sketch implementing Algorithm 4.1 and two real-world-based examples follow.

```
% INPUTS
c = input('Enter the objective function coefficient
vector (f=c´*x): ');
A = input('Enter the left-hand side coefficient matrix of linear
inequality constraints (A*x<=b): ');
b = input('Enter the right-hand side value vector of linear
inequality constraints (A*x<=b): ');

% INIT
n = length(c);
```

```
m = length(b);

if ~isempty(find(b<0,1))
   A(find(b<0,1),:) = -A(find(b<0,1),:);
   b(find(b<0,1)) = -b(find(b<0,1));
end

% CREATING THE INITIAL SIMPLEX TABLE
Table = [A, eye(m), b; -c', zeros(1,m), 0]

% ITERATIONS (= SIMPLEX TABLE TRANSFORMATIONS)
IsOptimal = isempty(find(Table(end,1:(n+m))<0,1));
while ~IsOptimal
    [c_min,pivot_col] = min(Table(end,1:(n+m))); % PIVOT COLUMN
   % PIVOT ROW (RATIO TEST)
   ratio_min = inf;
   for i = 1:1:m
      if (Table(i,end)/Table(i,pivot_col))>=0 &&
(Table(i,end)/Table(i,pivot_col))<ratio_min
         ratio_min = Table(i,end)/Table(i,pivot_col);
         pivot_row = i;
      end
   end
   if ratio_min == inf
      disp('A bounded optimal solution does not
exist.');
      return
   end
   % PIVOT ELEMENT ELIMINATION
   A_pivot_alt = Table(pivot_row,:)./
Table(pivot_row,pivot_col);
   Table(pivot_row,:) = A_pivot_alt;
   for i = 1:1:(m+1)
      if i ~= pivot_row
         Table(i,:) = Table(i,:)-A_pivot_alt*
Table(i,pivot_col);
      end
   end
   % OPTIMALITY TEST
   IsOptimal = isempty(find(Table(end,1:(n+m))<0,1));
   Table
end


% DETERMINING BASE VARIABLES, THEIR VALUES, AND SENSITIVITY ANAL-
YSIS
X = zeros((n+m),1);
ShadowPrices = [];
ReducedPrices = [];
k1 = 0; k2 =0;
for i = 1:1:(m+n)
   x = Table(:,i);
   if sum(x)==1 && min(x)==0 && max(x)==1 % BASE
```

```
VARIABLE
      ind_x = find(x==1);
      X(i) = Table(ind_x, end);
      if X(i)==0
        disp('A degenerative solution found.');
      end
    else % NON-BASE VARIBLE
      X(i) = 0;
      if x(end) == 0
        disp('Multiple optimal solution exist.');
      end
      if i<=n
        k1 = k1+1;
        ReducedPrices(k1,1:2) = [i, x(end)];
      else
        k2 = k2+1;
        ShadowPrices(k2,1:2) = [i, x(end)];
      end
    end
end

% OUTPUTS
% OPTIMAL FITNESS FUNCTION VALUE
str = sprintf("The optimal f = %0.5g", Table(end,end));
disp(str);
disp('The optimal variable values are:');
for i = 1:1:(m+n)
   str = sprintf('x%d = %0.5g', i, X(i,1));
   disp(str);
end
disp('The reduced prices are:');
for i = 1:1:k1
   str = sprintf('c%d = %0.5g', ReducedPrices(i,1),
ReducedPrices(i,2));
   disp(str);
end
disp('The shadow prices are:');
for i = 1:1:k2
   str = sprintf('b%d = %0.5g', ShadowPrices(i,1),
ShadowPrices(i,2));
   disp(str);
end
```

Example 4.3 *Task*: The company intends to manufacture five types of wooden and metal products at ZAR 15,000, 5000, 10,000, 10,000, and 30,000 per unit, respectively. It has a sufficient amount of metal material. Wood consumptions for producing a product unit are 2, 0, 1, 1, and 2 m^3, respectively, and storage spaces are limited to a capacity of 120 m^3/day. There are two processing facilities, and they have the following energy requirements: The 1st machinery: 0.5, 1, 2, 0, and 0 kWh/unit; 2nd device: 1, 2, 0, 1, and 0.5 kWh/unit. The management of the company decided that the maximum daily energy consumption (considering the electricity tariff and the machinery wear) for the first-type equipment will be 200 kWh and for the second-type machinery 400 kWh. Determine the optimal production schedule, i.e., how many units of which products should be produced daily to maximize sales under the given constraints. Analyze the solution.

Solution: Assemble the mathematical model first. Denote by x_1, x_2, x_3, x_4, x_5 units (pieces) of each product type. Then, the fitness function reads

$$f(\mathbf{x}) = 15{,}000x_1 + 5000x_2 + 10{,}000x_3 + 10{,}000x_4 + 30{,}000x_5$$

Assume thousands of ZAR taken for simplicity, i.e.,

$$f(\mathbf{x}) = 15x_1 + 5x_2 + 10x_3 + 10x_4 + 30x_5$$

Constraints to used wooden material can be expressed as

$$2x_1 + x_3 + x_4 + 2x_5 \leq 120$$

and those to the first and second machinery energy consumption

$$0.5x_1 + x_2 + 2x_3 \leq 200$$
$$x_1 + 2x_2 + x_4 + 0.5x_5 \leq 400$$

respectively.

Following step 1 of Algorithm 4.1, the equation (standard) canonical form is

$$2x_1 + x_3 + x_4 + 2x_5 + x_6 = 120$$
$$0.5x_1 + x_2 + 2x_3 + x_7 = 200$$
$$x_1 + 2x_2 + x_4 + 0.5x_5 + x_8 = 400$$

where $x_6, x_7, x_8 \geq 0$ represent slack variables. Now, it is possible to assemble the initial simplex table

x_1	x_2	x_3	x_4	x_5	x_6	x_7	x_8	b_i
2	0	1	1	2	1	0	0	120
0.5	1	2	0	0	0	1	0	200
1	2	0	1	0.5	0	0	1	400
– 15	– 5	– 10	– 10	– 30	0	0	0	0

The current solution is not optimal since a negative number in the last row exists (see step 3 of the algorithm). Hence, find the pivot column given by the "most negative" value in the last row as per (4.11), i.e.,

$$j^p = \arg\min(-15, -5, -10, -10, -30) = \arg(-30) = 5$$

The pivot row is determined by the ratio test (4.12)

$$i^p = \arg \min_{\geq 0}(120/2,200/0,400/0.5) = \arg(60) = 1$$

Hence, the pivot element equals 2 in the first row and the fifth column, as highlighted below

x_1	x_2	x_3	x_4	x_5	x_6	x_7	x_8	b_i
2	**0**	**1**	**1**	*2*	**1**	**0**	**0**	**120**
0.5	1	2	0	0	0	1	0	200
1	2	0	1	0.5	0	0	1	400
– 15	– 5	– 10	– 10	– 30	0	0	0	0

The pivot element can be eliminated, e.g., as follows: Multiply the pivot row by 0.5, add the (– 0.5)-multiple of this adjusted row to the 3rd row and its 30-multiple to the last row. After this transformation step, the eventual simplex table form reads

x_1	x_2	x_3	x_4	x_5	x_6	x_7	x_8	b_i
1	0	0.5	0.5	1	0.5	0	0	60
0.5	1	2	0	0	0	1	0	200
0.5	2	– 0.25	0.75	0	– 0.25	0	1	370
15	– 5	5	5	0	15	0	0	1800

Notice that the fitness function value increases to 1800 (thousands ZAR); however, the solution is not optimal yet as the last row includes the number – 5. This number determines the pivot column ($j^p = 2$), and the pivot row is

$$i^p = \min_{\geq 0}(60/0,200/1,370/2) = \arg(185) = 3$$

The pivot row can be multiplied by 0.5, and this altered row can further be subtracted from the 2nd row, and its 2.5-multiple be added to the last row. The eventual simplex table form reads

x_1	x_2	x_3	x_4	x_5	x_6	x_7	x_8	b_i
1	0	0.5	0.5	1	0.5	0	0	**60**
0.25	0	2.125	– 0.375	0	0.125	1	– 0.5	**15**
0.25	1	– 0.125	0.375	0	– 0.125	0	0.5	**185**
16.25	0	***4.875***	***6.875***	0	**14.375**	0	**2.5**	*2725*

that is optimal because all numbers in its last row (except for the rightmost column) are non-negative.

The results can be deduced as follows:

- The maximum objective function value is $f^* = 2725$ (italics-highlighted), which means that the daily sale is ZAR 2,725,000.
- The numbers of manufactured product units are determined by the positions of 1 in the underline-highlighted unit vectors that agree with the base variables. The corresponding values lie in the rightmost column (bold-highlighted). Non-basic variables given by the non-unit vectors equal zero. Hence

$$x_2^* = 185,\ x_5^* = 60,\ x_7^* = 15$$
$$x_1^* = x_3^* = x_4^* = x_6^* = x_8^* = 0$$

- It means, in practice, that products no. 1, 3, and 4 are not worth making.
- The non-zero value of slack variable x_7^* indicates that the second constraint (on the electricity consumption of the first-type machineries) is not exhausted, and only 185 kWh is consumed daily. Contrariwise, the second-type machineries consume all 400 kWh, and also, the whole available wooden material is exhausted. The arithmetic check can easily be done

$$f = 15 \cdot 0 + 5 \cdot 185 + 10 \cdot 0 + 10 \cdot 0 + 30 \cdot 60 = 2725$$

$$2 \cdot 0 + 0 + 0 + 2 \cdot 60 + 0 = 120$$
$$0.5 \cdot 0 + 185 + 2 \cdot 0 + 15 = 200$$
$$0 + 2 \cdot 185 + 0 + 0.5 \cdot 60 + 0 = 400$$

- The sensitivity analysis reveals whether it is reasonable to change the constraint bounds. The relevant values (i.e., shadow prices) are highlighted in bold underline. The value of 14.375 implies that whenever the storage capacities for the wooden material increase by 1 m^3, the daily sales increase by ZAR 14,375. If the company management allows the daily energy consumption of the second-type production equipment to increase by 1 kWh, sales will increase by ZAR 2500. Of course, economists and engineers have to take the data to judge. Note that the zero in the column below x_7 is natural since it makes no sense to increase the capacity if the particular resource is not exhausted.
- Similarly, the reduced prices can be determined from the bold italics-highlighted numbers. Value 16.25 means that a unit of product no. 1 should be for ZAR 15,000+16,250 = 31,250 to be worth selling. Analogously, for the third-type and fourth-type products, their units should be for ZAR 10,000 + 4875 = 14,875 and ZAR 10,000 + 6875 = 16,875, respectively, so the particular variable becomes basic.

4.2.3 *Simplex-Table Method for Mixed-Constraint LP Problem*

Various LP tasks include constraints of $=$ or $\geq$ type, not only $\leq$, see (4.3). These cases require specific operations when solving the problems using the simplex-table method. The so-called *two-step* simplex-table method is used. The first step verifies whether an LP problem solution exists by minimizing the auxiliary cost function. The second step then follows the standard simple-table method introduced in Sect. 4.2.2.

Assume the max $f(\mathbf{x})$ problem as in (4.1). Consider three different cases for which the following operations are made to assemble the initial canonical form of the simplex table:

- If the constraint is of $\leq$ type, a *slack* variable with the *plus* sign is introduced (as described in the preceding section).
- If the constraint is of $=$ type, an **artificial variable** with the *plus* sign is introduced. Artificial variables have a different meaning than the slack ones. They do not express a margin to the constraint, but they are inserted into the simplex table technically. The goal is to add the unit vector related to the particular constraint so that the basic solutions are obtained. Artificial variables *must be zero*; otherwise, the solution to the LP problem does not exist.
- If the constraint is of $\geq$ type, a *slack* variable with the *minus* sign expressing an "overdraft" of the constraint is introduced. However, this operation does not insert the unit vector into the simplex table. Hence, an *artificial* variable (with the plus sign) must also be added.

The auxiliary function h to be minimized is constructed as a sum of all artificial variables (with the plus sign). The outcome of the minimization must be that $h\left(\mathbf{x}_a^*\right) = 0$ with $\mathbf{x}_a^* = \mathbf{0}_{n_a}$, where $\mathbf{x}_a^*$ is the vector of all n_a artificial variables. Any other optimal solution implies that the given LP problem does not have a solution (i.e., there is no intersection of the given constraints).

The minimization procedure is summarized in Algorithm 4.2.

Algorithm 4.2 *(The minimization of the auxiliary function in the two-step simplex-table method)*

1. Consider fitness function (4.1) and mixed linear constraints (equalities and/or inequalities). Based on particular constraints, construct the initial canonical simplex table form as described above. The pre-last row is composed of $-\mathbf{c}^T$ and followed by zeros in the remaining columns as in step 2 of Algorithm 4.2. The last row is given by $h(\mathbf{x}_a) = \sum_{i=1}^{n_a} x_{a,i}$, so the number -1 is placed in the columns under each artificial variable. The rightmost column includes zero again.
2. Add all rows that include artificial variables into the last row.
3. If all coefficients in the last row (except for the rightmost column) are *non-positive*, the minimization is optimal, and go to step 7. Else, go to step 4.

4. Find the pivot column, the index of which satisfies

$$j^p := \arg\max_j \{\text{numbers in the last row}\} \tag{4.13}$$

5. The pivot row is determined the accordance with the ratio test (4.12). Then, the corresponding intersection gives rise to the pivot element.
6. As per step 6 of Algorithm 4.1, the pivot element is eliminated. Then, go to step 3.
7. The basic and non-basic variablesare deduced in accordance to step 7 of Algorithm 4.1. If $\exists x^*_{a,i} \in \mathbf{x}^*_a$ that is a basic variable, the LP problem does not have a solution; else, cancel the last row of the simplex table and continue with step 4 of Algorithm 4.1 for maximization of $f(\mathbf{x})$.

Some remarks to Algorithm 4.2 follow:

- When assembling the canonical form in step 1 of the algorithm, if $\exists b_i < 0, i = 1, 2, \ldots, m$, the corresponding (in)equality is multiplied by -1.
- Step 2 of the algorithm means that all artificial variables become basic. If, after this step, the solution is not optimal (see step 3), other steps attempt to eliminate the artificial variables from the basis. Whenever the optimal solution is reached and no artificial variable is basic, it means that a solution to the given mixed-constraint LP problem exists (and it is found via Algorithm 4.1). Otherwise, if artificial variable remains basic, the constraints have no intersection.
- When continuing with Algorithm 4.1 after step 7 of Algorithm 4.2, the columns related to the artificial variables can be omitted, as these variables are already known to equal zero. However, if they remain in the simplex table, they can be ignored when performing the optimality test as per step 3 of Algorithm 4.1.

An illustrative example follows.

Example 4.4 *Task*: Consider the task assignment as in Example 4.3. However, let

the energy consumption be *exactly* 200 kWh and 400 kWh, respectively. How will the solution differ? Will the fitness function increase or decrease?

Solution: Recall that the first-type machineries consumed only 185 kWh out of the allowable 200 kWh, so there was a margin to the constraint in the optimal solution of Example 4.3. Intuitively, higher energy consumption might yield better profits. Assemble the mathematical model. The fitness function $f(\mathbf{x})$ is identical to that presented in Example 4.3. However, the constraint set differs as follows

$$\begin{aligned} 2x_1 + x_3 + x_4 + 2x_5 &\leq 120 \\ 0.5x_1 + x_2 + 2x_3 &= 200 \\ x_1 + 2x_2 + x_4 + 0.5x_5 &= 400 \end{aligned}$$

i.e., there are two equalities instead of inequalities.

The standard (canonical) form is obtained by adding a slack variable with the plus sign for the inequality of the $\leq$ type (in accordance with the preceding example); however, the equality constraints require introducing an artificial variable (with the plus sign) for each constraint. The canonical for then reads

$$
\begin{aligned}
2x_1 + x_3 + x_4 + 2x_5 + x_6 &= 120 \\
0.5x_1 + x_2 + 2x_3 + x_{a,1} &= 200 \\
x_1 + 2x_2 + x_4 + 0.5x_5 + x_{a,2} &= 400
\end{aligned}
$$

where x_6 is a slack variable and $x_{a,1}, x_{a,2}$ are artificial variables.

When solving a mixed-constraint LP problem, verifying that all the artificial variables are zero is first necessary. That is, these variables should be non-basic. In this example, the verification is done via the minimization of the auxiliary cost function

$$\min h\left(x_{a,1}, x_{a,2}\right) = x_{a,1} + x_{a,2}$$

Following step 1 of Algorithm 4.2, the initial canonical simplex table form is

x_1	x_2	x_3	x_4	x_5	x_6	$x_{a,1}$	$x_{a,2}$	b_i
2	0	1	1	2	1	0	0	120
0.5	1	2	0	0	0	1	0	200
1	2	0	1	0.5	0	0	1	400
– 15	– 5	– 10	– 10	– 30	0	0	0	0
0	0	0	0	0	0	– 1	– 1	0

Step 2 of the algorithm commands to add the 2nd and 3rd rows to the last one so that the artificial variables become basic, i.e.,

x_1	x_2	x_3	x_4	x_5	x_6	$x_{a,1}$	$x_{a,2}$	b_i
2	0	1	1	2	1	0	0	120
0.5	1	2	0	0	0	1	0	200
1	2	**0**	**1**	**0.5**	**0**	**0**	**1**	**400**
– 15	– 5	– 10	– 10	– 30	0	0	0	0
1.5	3	2	1	0.5	0	0	0	600

One can deduce the current results from the table: $x_1 = x_2 = x_3 = x_4 = x_5 = 0$, $x_6 = 120$, $x_{a,1} = 200$, $x_{a,2} = 400, f(\mathbf{x}) = 0$ (see step 7 of Algorithm 4.1). However, the minimization of $h(\mathbf{x}_a)$ is not finished as a positive number exists in the last row (see step 3 of Algorithm 4.2). Now, the pivot column index is determined by (4.13) as $j^p = \arg\max(1.5,3,2,1) = \arg\max(3) = 2$, see the column highlighted in underline (Consider the equality $\min h = -\max(-h)$ for this step). The pivot row is determined by $i^p = \arg\min_{\geq 0}(120/0{,}200/1{,}400/2) = \arg(200) = 2$ or 3, see step 5 of Algorithms 4.1 and 4.2. Without loss of generality, let the 3rd row be selected (highlighted in bold). Then, the pivot element (highlighted in italics) is eliminated as per step 3 of Algorithm 4.1. After this procedure, the updated simplex table reads

x_1	x_2	x_3	x_4	x_5	x_6	$x_{a,1}$	$x_{a,2}$	b_i
2	0	1	1	2	1	0	0	120
0	**0**	*2*	**– 0.5**	**– 0.25**	**0**	**1**	**– 0.5**	**0**
0.5	1	0	0.5	0.25	0	0	0.5	200
– 12.5	0	– 10	– 7.5	– 28.75	0	0	0.25	1000
0	0	2	– 0.5	– 0.25	0	0	– 1.5	0

The table is not optimal with respect to minimizing $h(\mathbf{x}_a)$ (see the number 2 in the last row), and steps 4–6 of Algorithm 4.2 need to be proceeded again. Then, the table has the following form

x_1	x_2	x_3	x_4	x_5	x_6	$x_{a,1}$	$x_{a,2}$	b_i
2	0	0	1.25	2.125	1	– 0.5	0.25	120
0	0	1	– 0.25	– 0.125	0	0.5	– 0.25	0
0.5	1	0	0.5	0.25	0	0	0.5	200
– 12.5	0	0	– 10	– 30	0	5	0	1000
~~0~~	~~0~~	~~0~~	~~0~~	~~0~~	~~0~~	~~– 1~~	~~– 1~~	~~0~~

Now, the minimization of $h(\mathbf{x}_a)$ is finished as per step 3 of Algorithm 4.2, and eliminating the artificial variables for the basis is successful, i.e., $x_{a,1}^* = x_{a,2}^* = 0$ with $h(\mathbf{x}_a^*) = 0$. It means that the LP problem is solvable, and the last row of the simplex table can be removed. Notice also that while the minimization procedure has been performed, the maximization of $f(\mathbf{x})$ has been simultaneously solved. Thus, the current fitness function value is $f = 1000$; however, it is not optimal due to the existence of negative values in the last row of the updated table under the decision and slack variables.

The second step of the two-step simplex-table method then follows Algorithm 4.1. The eventual simplex table reads

x_1	x_2	x_3	x_4	x_5	x_6	$x_{a,1}$	$x_{a,2}$	b_i
0.94	0	0	0.59	1	0.47	– 0.235	0.117	56.47
0.118	0	1	– 0.18	0	0.06	0.47	– 0.235	7.06
0.265	1	0	0.35	0	– 0.12	0.06	0.471	185.88
15.73	0	0	7.65	0	14.12	– 2.06	3.53	2697.12

The table is optimal regardless of the negative value under $x_{a,1}$ (highlighted in underline). It has already been verified in the first step of the two-step method that

$x_{a,1}^* = 0$; therefore, the particular column cannot be pivoted (and hence, $x_{a,1}$ become basic).

Hence, the optimal solution is

$$f(\mathbf{x}^*) = 2697.12$$
$$x_2^* = 185.88, x_3^* = 7.06, x_5^* = 56.47$$
$$x_1^* = x_4^* = x_6^* = 0$$

Compared to the results of Example 4.3, the optimal fitness function value decreased, i.e., the daily sales became worse. It seems paradoxical at first sight since the overall electricity consumption is higher than in the preceding task. However, it generally holds that the stricter constraints yield worse results (e.g. when insisting on exhausting the resources).

As the resulting optimal units to be produced are non-integer, there is no unique answer about the production scheme. Generally, the optimal integer values cannot be obtained by simply rounding a non-integer result (see Sect. 4.3).

4.2.4 Primal and Dual LP Problem

In this section, the so-called *symmetric* **dual LP problem** is touched. For other, more advanced types of LP dual problems, readers are referred to provided literature resources [6, 9, 11].

For each **primal** LP problem, a corresponding **dual** LP problem can be assigned. Let the primal model be the basic LP problem defined in (4.1)–(4.2). Then, the symmetric dual problem reads

$$\min g(\mathbf{y}) = b_1 y_1 + b_2 y_2 + \cdots + b_m y_m = \mathbf{b}^T \mathbf{y} \in \mathrm{R}, \mathbf{y} \in \mathrm{R}^m$$
$$\left.\begin{array}{l} a_{11} y_1 + a_{21} y_2 + \cdots + a_{m1} y_m \geq c_1 \\ a_{12} y_1 + a_{22} y_2 + \cdots + a_{m2} y_m \geq c_2 \\ \ldots \\ a_{1n} y_1 + a_{2n} y_2 + \cdots + a_{mn} y_m \geq c_n \end{array}\right\} \Leftrightarrow \mathbf{A}^T \mathbf{y} \geq \mathbf{c} \tag{4.14}$$
$$y_i \geq 0, i = 1, 2, \ldots, m$$

The following basic statements hold for the primal and dual models.

Theorem 4.2 (Basic theorem about the finite value) *Suppose one of the models (primal or dual) has a solution with a finite objective function value. In that case, the second (corresponding) model has a solution with a finite objective function value, and both the objective functions are equal (which also holds in the case of an optimal solution).*

$$\max f(\mathbf{x}) = \min g(\mathbf{y}) \tag{4.15}$$

Otherwise, if one of the models does not have a finite solution, the other one does not have an admissible solution.

Theorem 4.3 *If one of the models (primal or dual) has more optimal solutions, the solution of the second (corresponding) model is degenerated.*

Note that the physical dimension of y_i can easily be deduced from the constraint set and dimensions of a_{ij}, c_i.

The optimal solution to problem (4.14) can be done by transforming this model to (4.1), which is solved by Algorithm 4.1. Then, the optimal solution of (4.14) can be deduced from the final form of the simplex table as the values of in the last row related to (i.e., below) slack variables. The following example demonstrates this procedure.

Example 4.5 *Task*: The mining company mines three ores in a location: gold, palladium, and silver. As the working environment is hazardous to health and infested with chemicals, it is the union's effort and demand to minimize the time company employees spend in the mines. Mining one ton of gold ore takes 6 min, as it is in an inaccessible and already heavily mined area; palladium takes only 3 min, and finally, a ton of silver ore takes only 72 s. The extractions of the precious metal from the ore are as follows: 0.1 g of pure gold is obtained from 1 kg of gold ore, 0.3 g of palladium is obtained from 1 kg of palladium ore, and for silver is 0.6 g of metal per 1 kg of ore. The net profits from ore mining for gold, palladium, and silver, respectively, are \$5000, \$1000, and \$200 per 1 kg of metal. Simultaneously, the company must deliver at least 2 kg of gold, 10 kg of palladium, and 80 kg of silver daily. A minimum daily operating profit of \$50,000 is also required. Determine the company's mining plan and the time employees stay there per shift and analyze the result.

Solution: Assemble a mathematical problem formulation, including the physical dimensions of the values and variables. Apparently, the goal is to minimize the mining time (duration), which can be expressed by

$$6\,[\text{min}/\text{ton}] \cdot x_1\,[\text{ton}] + 3\,[\text{min}/\text{ton}] \cdot x_2\,[\text{ton}] + 1.2\,[\text{min}/\text{ton}] \cdot x_3\,[\text{ton}]$$

where x_1, x_2, x_3 are tons of mined gold, palladium, and silver ores, respectively. The result has the dimension of minutes (min). The same cost function can also be formulated in hours (h) and tens of tons:

$$\begin{aligned} 1\,[\text{h}/(10\cdot\text{ton})] \cdot x_1\,[10\cdot\text{ton}] &+ 0.5\,[\text{h}/(10\cdot\text{ton})] \cdot x_2\,[10\cdot\text{ton}] \\ &+ 0.2\,[\text{h}/(10\cdot\text{ton})] \cdot x_3\,[10\cdot\text{ton}] \end{aligned}$$

Hence, the eventual cost function reads

$$\min f(x_1, x_2, x_3) = x_1 + 0.5\,x_2 + 0.2\,x_3\,[\text{h}]$$

The constraints are given by the requirements for the minimum daily profit and necessary metal deliveries. If all variables are in tens of tons, the required mass of gold is expressed by

$$1\left[\text{kg}/(10\cdot\text{ton})\right]\cdot x_1\,[10\cdot\text{ton}]\geq 2\left[\text{kg}\right]$$

Analogously, it should hold for masses of pure palladium and silver metal, respective:

$$\begin{aligned}3\left[\text{kg}/(10\cdot\text{ton})\right]\cdot x_2\,[10\cdot\text{ton}]&\geq 10\left[\text{kg}\right]\\6\left[\text{kg}/(10\cdot\text{ton})\right]\cdot x_3\,[10\cdot\text{ton}]&\geq 80\left[\text{kg}\right]\end{aligned}$$

Finally, the constraint on the minimum required daily profit is obtained by analyzing of the extraction versus net profits:

$$\begin{aligned}&5000\left[\$/\text{kg}\right]\cdot 1\left[\text{kg}/(10\cdot\text{ton})\right]\cdot x_1\,[10\cdot\text{ton}]\\&\quad+1000\left[\$/\text{kg}\right]\cdot 3\left[\text{kg}/(10\cdot\text{ton})\right]\cdot x_2\,[10\cdot\text{ton}]\\&\quad+200\left[\$/\text{kg}\right]\cdot 6\left[\text{kg}/(10\cdot\text{ton})\right]\cdot x_3\,[10\cdot\text{ton}]\\&\geq 50{,}000\left[\$\right]\end{aligned}$$

Hence, the eventual optimization problem (after some simple algebraic manipulation) reads

$$\begin{aligned}&\min f(x_1,x_2,x_3)=x_1+0.5\,x_2+0.2\,x_3\\&x_1\geq 2\\&x_2\geq 10/3\\&x_3\geq 40/3\end{aligned}$$

i.e.,

$$\mathbf{c}=[1,\ 0.5,\ 0.2]$$
$$\mathbf{b}=[2,\ 10/3,\ 40/3,\ 50]^T$$
$$\mathbf{A}=\begin{bmatrix}1&0&0\\0&1&0\\0&0&1\\5&3&1.2\end{bmatrix}$$

Apparently, this model is dual to the basic (primary) LP problem (4.1)–(4.2). Hence, the associated (primal) model that can be solved by the simplex-table method (see Algorithm 4.1) is formulated as follows

$$\max\ g(y_1, y_2, y_3, y_4) = \mathbf{b}^T\mathbf{y} = 2y_1 + 10/3y_2 + 40/3y_3 + 50y_4$$

$$\mathbf{A}^T\mathbf{y} \le \mathbf{c}^T \Leftrightarrow \begin{bmatrix} 1 & 0 & 0 & 5 \\ 0 & 1 & 0 & 3 \\ 0 & 0 & 1 & 1.2 \end{bmatrix} \begin{bmatrix} y_1 \\ y_2 \\ y_3 \\ y_4 \end{bmatrix} \le \begin{bmatrix} 1 \\ 0.5 \\ 0.2 \end{bmatrix}$$

$$\Leftrightarrow$$

$$\begin{aligned} y_1 + 5y_4 &\le 1 \\ y_2 + 3y_4 &\le 0.5 \\ y_3 + 1.2y_4 &\le 0.2 \end{aligned}$$

Note that this primal model contains the identical ratio $(0.5/3 = 0.2/1.2)$ in the second and third nonequality condition for y_4. It yields a (partially) degenerated solution, as two polyhedron vertices in $\mathbb{R}^4$ coincide; see the comments on Algorithm 4.1.

It is also interesting to analyze the physical dimension of y_i. The right-hand side of the first constraint has the dimension of $1\,[\mathrm{h}/(10 \cdot \mathrm{ton})]$ and the left-hand one $1\left[\mathrm{kg}/(10 \cdot \mathrm{ton})\right]$; therefore, y_i has the dimension of $\left[\mathrm{h}/\mathrm{kg}\right]$.

First, the inequalities (constraints) are transformed to the canonical equation form

$$\begin{aligned} y_1 + 5y_4 + y_5 &= 1 \\ y_2 + 3y_4 + y_6 &= 0.5 \\ y_3 + 1.2y_4 + y_7 &= 0.2 \end{aligned}$$

where y_5, y_6, y_7 are slack variables. Then, the simple table method gives

y_1	y_2	y_3	y_4	y_5	y_6	y_7	b_i
1	0	0	5	1	0	0	1
0	1	0	*3*	0	1	0	0.5
0	0	1	1.2	0	0	1	0.2
– 2	– 10/3	– 40/3	– 50	0	0	0	0

y_1	y_2	y_3	y_4	y_5	y_6	y_7	b_i
1	– 5/3	0	0	1	– 5/3	0	1/6
0	1/3	0	1	0	1/3	0	1/6
0	– 2/5	*1*	0	0	– 2/5	1	0
– 2	40/3	– 40/3	0	0	50/3	0	50/6

y_1	y_2	y_3	y_4	y_5	y_6	y_7	b_i
1	– 5/3	0	0	1	– 5/3	0	1/6
0	1/3	0	1	0	1/3	0	1/6
0	– 2/5	1	0	0	– 2/5	1	0
– 2	8	0	0	0	34/3	40/3	50/6

y_1	y_2	y_3	y_4	y_5	y_6	y_7	b_i
1	– 5/3	0	0	1	– 5/3	0	1/6
0	1/3	0	1	0	1/3	0	1/6
0	– 2/5	1	0	0	– 2/5	1	0
0	14/3	0	0	***2***	***8***	***40/3***	26/3

The pivot element at each step is highlighted in italics. The optimal solution of the primal problem is: $y_1 = y_4 = 1/6$, $y_3 = 0$, $y_2 = y_5 = y_6 = y_7 = 0$. The degenerated solution is indicated by $y_3 = 0$ that is basic. Another optimization procedure can be obtained by selecting 1.2 as the pivot element in the first iteration step. Recall that the relation between a degenerated solution and an infinite number of solutions between the models is formulated in Theorem 4.3.

Nevertheless, the intention is not to get the solution for the primary LP problem but for the dual model (i.e., x_i). This optimal solution can be deduced from the last line below the slack variables (highlighted in bold italics). Hence, $x_1 = 2, x_2 = 8, x_3 = 40/3$, and according to Theorem 4.2, it holds that

$$\min f(\mathbf{x}) = \max g(\mathbf{y}) = 26/3 = 8\frac{2}{3}$$

The solution interpretation is as follows: The company mines 20 tons of gold ore, 40 tons of palladium ore, and 133 and 1/3 tons of silver ore daily, which takes 8 h and 40 min. Verify the cost function value and validity of the constraints:

$$\begin{aligned}
&f = 2 + 0.5 \cdot 8 + 0.2 \cdot 40/3 = 26/3 \\
&2 = 2 \\
&8 \geq 10/3 \\
&40/3 = 40/3 \\
&5 \cdot 2 + 3 \cdot 8 + 1.2 \cdot 40/3 = 50
\end{aligned}$$

Notice that the requirement of the daily delivery demand on palladium metal is the only constraint that remains in the form of inequality. Hence, $8 - 10/3 = 14/3$ tens of tons of silver ore are mined more than necessary. This constraint margin (overlap) can also be found in the simplex table below y_2 (highlighted in underline).

4.3 Integer LP Problem

The integer LP (ILP) task means finding an optimal (non-negative) integer solution to the given LP problem, i.e., $\mathbf{x} \in \mathbb{Z}_+$ [17]. The important fact is that if the optimal solution to an LP problem is an integer, it is also the optimal solution of the ILP task. Contrariwise, if the optimal solution is a non-integer, the optimal solution of the ILP problem has to be further found (e.g., by particular techniques based on the simplex table). It must be stressed that rounding the non-integer solution does not yield the optimal integer solution in general.

Figure 4.6 demonstrates the positions of possible ILP problem solutions in $\mathbb{R}^+$, which can be imagined as a net of discrete points (nodes).

In the following sections, two computational methods to solve an ILP problem are introduced.

4.3.1 *Branch-and-Bound Method*

The framework of this method for general discrete programming problems was designed by Ailsa Land (1927–2021) and Alison Harcourt (2029) in 1960. The principle of the method can be summarized as follows. It starts with finding a possible non-integer optimal LP solution via Algorithm 4.1 or Algorithm 4.2. If the obtained solution is non-integer, then the remaining procedure is based on the decomposition of the constraint set to two disjunctive subsets. A basic variable x_i of a non-integer value $y_i = x_i$ is selected, and the following constraints supplement the constraint set:

$$\begin{aligned} x_i &\leq \lfloor y_i \rfloor \\ x_i &\geq \lfloor y_i \rfloor + 1 \end{aligned} \tag{4.16}$$

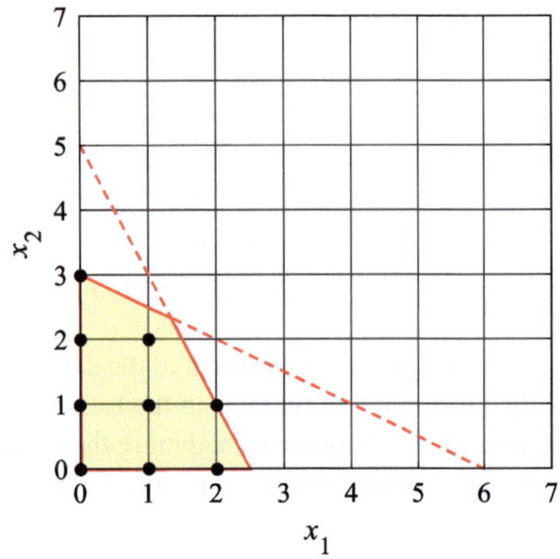

Fig. 4.6 Possible solutions to an ILP problem

where $\lfloor\cdot\rfloor$ denotes the floor function (i.e., the nearest lower integer). Both the subproblems are solved separately. This operation represents the solution "branching", and this computation is repeated in the other two subbranches if necessary. It is solved either via a recursive computing or a queue of unsolved subproblems. The branching is terminated whenever the found solution is non-admissible (i.e., out of the given current constraint set) or it is an integer one. Hence, the solution tree reaches its leaves—"bounding".

As any integer solution cannot beat a non-integer one for the given problem, the branching also stops in cases when the found non-integer solution is worse than the current best integer one.

Whenever a particular new condition (4.16) is added to the subtask, it is not necessary to resolve the whole simplex table again. The resolving can be done by applying a step of the so-called **dual method**, which can be done for fitness function (4.1) as follows:

Algorithm 4.3 *(One step of the dual method)*

1. A new condition (4.16) is transformed to the standard form (see step 2 of Algorithm 4.1 or Sect. 4.2.3), and it is multiplied by -1. The particular coefficients a_{dj_d} are added to the simplex table as a row at the end of constraints (i.e., d row).
2. This newly added row $i^p = d$ (or, generally, that satisfying $i^p := \arg\min_i\{\text{numbers in the last column} < 0\}$) is taken as the *pivot row*.
3. The *pivot column* is found as

$$j^p := \arg\max_j \left\{ \gamma_j = \frac{c_j}{a_{i^p j}} < 0 \right\} \tag{4.17}$$

where c_j are coefficients in the last row of the simplex table (i.e., those of the fitness function), except for the right-most column.
4. The *pivot element* at position $\left[i^p, j^p\right]$ is eliminated by the standard row operations as in step 6 of Algorithm 4.1.

Considering objective function (4.1), the branch and bound method can be sketched by Algorithm 4.4.

Algorithm 4.4 *(The branch-and-bound method)*

1. Let an admissible optimal solution to the ILP problem obtained via Algorithm 4.1 be $\mathbf{x}_R^* \in \mathbb{R}^{n+m}$, where subscript R denotes a real-valued (i.e., a non-integer) solution.
2. The "best" found integer solution (indicated by subscript Z) and its fitness function value are initially set as $f_Z^* := -\infty$.
3. An arbitrary non-integer basic variable $x_i \in \mathbf{x}_R^*$, $x_i \notin \mathbb{Z}$, $i = 1, 2, \ldots, n$ of value $y_i = x_i$ is selected (usually, that with the highest non-integer part), and two distinct conditions (4.16) are added to the two corresponding constraint lists using Algorithm 4.3:

(a) $x_i \leq \lfloor y_i \rfloor$, and the corresponding LP problem is solved via the simplex table, yielding result $\mathbf{x}_1^*, f_1 = f(\mathbf{x}_1^*)$; and
(b) $x_i \geq \lfloor y_i \rfloor + 1$, and the corresponding LP problem is solved via the simplex table, yielding result $\mathbf{x}_2^*, f_2 = f(\mathbf{x}_2^*)$.

4. For each of the two branches in step 4, it is performed a test whether the found solution is admissible:

   (a) If it is admissible, go to step 5; else,
   (b) This branch is terminated (i.e., a bound), and return from recursion to step 4.

5. If the solution is admissible (for each of the branches), a test of whether the found function value is higher than that of the best integer solution so far is performed:

   (a) If $f_i > f_Z^*, i \in \{1, 2\}$, go to step 6; else,
   (b) This branch is terminated (i.e., a bound), and return from recursion to step 4.

6.

   (a) If $x_j \in \mathbf{x}_i^* \in \mathbb{Z}, \forall j = 1, 2, \ldots, n$, set $f_Z^* = f_i, \mathbf{x}_Z^* = \mathbf{x}_i^*, i \in \{1, 2\}$. This branch is terminated (i.e., a bound), and return from recursion to step 4. Otherwise,
   (b) go to step 3 (into a deeper recursion or put ILP problems (4.16) into a queue of unsolved tasks).

Hence, the output of the recursive (or queue-based) procedure is the optimal value of the fitness (objective) function $f_Z^* \in \mathbb{R}$ and the corresponding integer solution vector $\mathbf{x}_Z^* \in \mathbb{Z}$. Alternatively, if test $f_i \geq f_Z^*$ is used in step 5 of Algorithm 4.4, more optimal solutions can possibly be found.

Example 4.6 *Task*: Find an optimal solution to the ILP problem

$$\max f(x_1, x_2) = 2x_1 + x_2$$

s.t.

$$\begin{aligned} x_1 + x_2 &\leq 5 \\ -x_1 + x_2 &\leq 0 \\ 6x_1 + 2x_2 &\leq 21 \\ x_1, x_2 &\in \mathbb{Z} > 0 \end{aligned}$$

using the branch-and-bound method.

Solution: The solution to the problem will only be indicated as a framework using the techniques mentioned in Examples 4.3 and 4.4 (i.e., the simplex-table method), as it is computationally demanding and lengthy.

Follow the branch-and-bound algorithm (Algorithm 4.4). First, an optimal admissible (non-integer) solution is found

x_1	x_2	x_3	x_4	x_5	b_i
1	1	1	0	0	5
− 1	1	0	1	0	0
6	2	0	0	1	21
− 2	− 1	0	0	0	0

x_1	x_2	x_3	x_4	x_5	b_i
0	*2/3*	1	0	− 1/6	3/2
0	4/3	0	1	1/6	7/2
1	1/3	0	0	1/6	7/2
0	− 1/3	0	0	1/3	7

x_1	x_2	x_3	x_4	x_5	b_i
0	1	3/2	0	− 1/4	9/4
0	0	− 2	1	1/2	1/2
1	0	− 1/2	0	1/4	11/4
0	0	1/2	0	1/4	31/4

Note that x_3, x_4, x_5 represent slack variables, and the pivot elements are italics-highlighted. The optimal non-integer solution then reads:

$$
\begin{aligned}
& f\left(\mathbf{x}_R^*\right) - 31/4 \\
& x_1 = 11/4 = 2.75,\ \ x_2 = 9/4 = 2.25, \\
& x_4 = 1/2,\ \ x_3 = x_5 = 0
\end{aligned}
$$

Now, set $f_Z^* := -\infty$ according to step 2 of Algorithm 4.4, and branch the algorithm into two subbranches as per (4.16). Consider, e.g., x_1 giving rise to

(a) $x_1 \leq 2$
(b) $x_1 \geq 3$.

Each of the conditions is added to the simplex table using Algorithm 4.3. For instance, condition (a) is added as follows

$$
\begin{aligned}
& x_1 \leq 2 \\
& \quad \Rightarrow x_1 + x_6 = 2 \\
& \quad \Rightarrow -x_1 - x_6 = -2
\end{aligned}
$$

where x_6 represents a slack variable, and the last equality is added to the table. This branch results in

$$\mathbf{x}^*_{Z,11} = [x_1, x_2] = [2, 2]$$
$$f_{11} = f\left(\mathbf{x}^*_{Z,11}\right) = 6$$

As this result is admissible and integer, then set $f^*_Z = f_{11} = 6$, $\mathbf{x}^*_Z = \mathbf{x}^*_{Z,11} = [2, 2]$ (according to step 6 of Algorithm 4.4), and this branch is terminated (i.e., a bound appears).

On the contrary, condition $x_1 \geq 3$ results in

$$\mathbf{x}^*_{R,12} = [x_1, x_2] = [3, 1.5]$$
$$f_{12} = f\left(\mathbf{x}^*_{R,12}\right) = 7.5$$

This solution is admissible, yet non-integer one. Since $f_{12} > f^*_Z$, it is reasonable to continue this branch by introducing two other disjunct inequality conditions

(a) $x_2 \leq 1$
(b) $x_2 \geq 2$

that are solved either as a recursive procedure or by adding them to queue of unsolved subtasks. The remaining solution tree is displayed in Fig. 4.7.

Hence, the optimal integer solution is

$$\mathbf{x}^*_Z = [3, 1], f^*_Z = 7$$

Notice that a simple rounding of the initial non-integer optimal LP problem solution does not lead to the optimal integer one.

The given task can also be solved graphically (see details in Sect. 4.2.1), as demonstrated in Fig. 4.8.

4.3.2 Gomory Method

This method, also called the Gomory cuts or the *cutting-plane* method, was investigated by Ralph Edward Gomory (1929). The method is based on a different philosophy than the branch-and-bound method; however, it also relies on a repetitive solution of ILP subproblems using the simplex-table method. In each iteration step, a new condition (constraint) is added via Algorithm 4.3. The recursion is not needed. The procedure continues until an (optimal) integer solution is found.

A possible disadvantage of the method is that it may converge slowly in some cases. Contrariwise, if a numerical error is made when computing, it may abruptly reach an integer solution that is, however, non-optimal (i.e., the optimal solution is "jumped over").

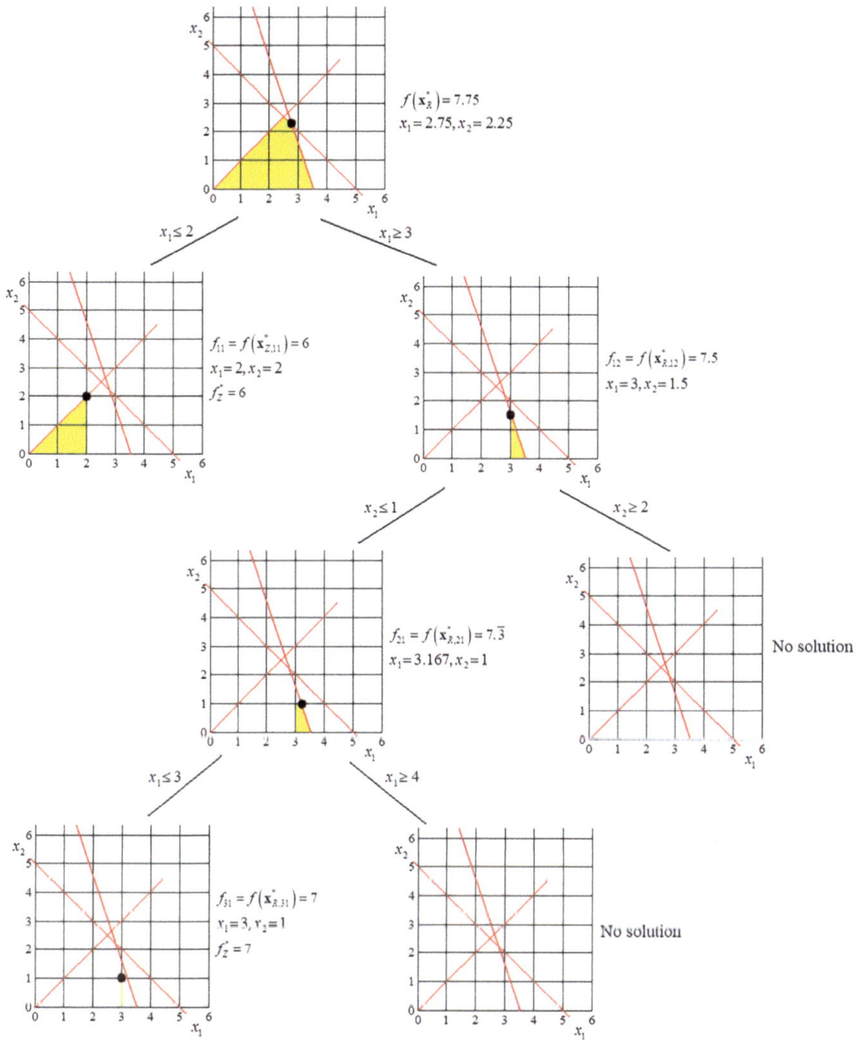

Fig. 4.7 The solution tree to Example 4.6

When assembling an additional constraint, the essential idea can be comprehended as the "imposing" the non-integer part of the solution upon all non-basic variables.

The procedure is summarized in Algorithm 4.5.

Algorithm 4.5 *(The Gomory method)*

1. Let an admissible optimal solution to the ILP problem obtained via Algorithm 4.1 be $\mathbf{x}_R^* \in \mathbb{R}^{n+m}$. Set $i = 1$ (the iteration counter).
2. Denote by

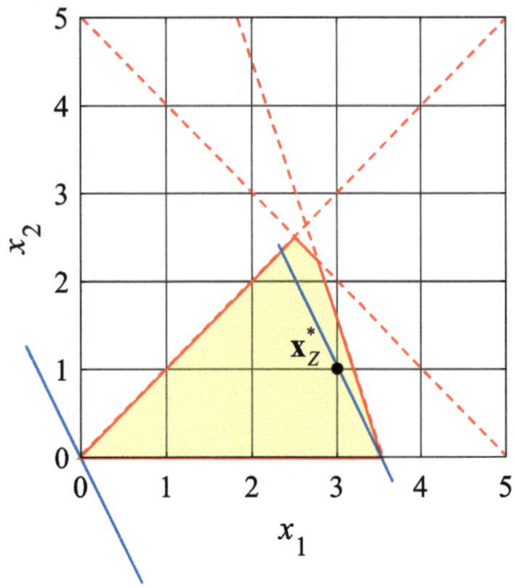

Fig. 4.8 The graphical solution to Example 4.6

$$x_j = \lfloor x_j \rfloor + \{x_j\}, j = 1, 2, \ldots n + m + i - 1 \tag{4.18}$$

the values of non-integer variables where $\lfloor x_j \rfloor$ means the floor function and $0 < \{x_j\} < 1$ expresses the corresponding non-integer reminder. Values (4.18) agree with coefficients in the right-most column for basic variables (see step 7 of Algorithm 4.1). Let the determination of the value of basic variable x_j corresponds to the kth row of the simple table, i.e., $x_j = b_k$ and vector $[0, 0, \ldots 0, 1, 0, \ldots 0]^T$ with one at the kth position appears below x_j.

Moreover, let constraint coefficients in the kth row be divided into the integer part and the non-integer reminder

$$a_{kj} = \lfloor a_{kj} \rfloor + \{a_{kj}\}, j = 1, 2, \ldots n + m + i - 1 \tag{4.19}$$

3. Find a basic non-integer variable $x_j \notin \mathbb{Z}$ with the highest non-integer reminder

$$j_{\max} = \arg\max\{x_j\}, x_j \notin \mathbb{Z}, j = 1, 2, \ldots, n \tag{4.20}$$

and the corresponding $k_{\max}$th row, i.e., $x_{j_{\max}} = b_{k_{\max}}$.

4. Introduce a new condition (constraint)—the so-called *Gomory's cut*:

$$\sum_{i=1}^{m+n+i-1} \{\alpha_{k_{\max},j}\}x_j - x_{s,i} = \{b_{k_{\max}}\} \tag{4.21}$$

where

$\{\alpha_{k_{\max},j}\} = 0$ if x_j is basic; else,
$\{\alpha_{k_{\max},j}\} = \{a_{k_{\max},j}\}$ if $a_{k_{\max},j} \geq 0$ for non-basic x_j; else,

$\{\alpha_{k_{\max},j}\} = a_{k_{\max},j} + z$ if $a_{k_{\max},j} < 0$ for non-basic x_j where $z > 0 \in \mathbb{Z}$ is the smallest positive integer such that $a_{k_{\max},j} + z > 0$.

The newly added variable $x_{s,i} \in \mathbb{Z}$ expresses an integer "margin".

5. Condition (constraint) (4.21) is multiplied by -1, and it is added to the simplex table (e.g., as a row above the objective function row). Then, the dual method is used starting from step 2 of Algorithm 4.3.
6. If the obtained solution is an integer ($\mathbf{x}_{\mathbb{Z}}^{*} \in \mathbb{Z}^{n+m+i-1}$), stop; else, set $i = i + 1$, and go to step 2.

Two demonstrative examples follow, and the latter is amusing.

Example 4.7 *Task*: The company sells two types of products. The first product is entirely plastic and is intended exclusively for a metallurgy company. 4 q of plastic granules are needed for production. The company receives EUR 1000 and 3 q of steel material from the metallurgy company. The second product is freely available for EUR 2000, and its production requires 3 q of plastic granules and 4 q of steel material. The plant-wide holiday is approaching, and they have 6 q of steel material and 12 q of plastic granules left in stock. The task is to plan production with the greatest possible cash sales without purchasing material for storage (except for deliveries of metallurgical steel). Solve the problem using the Gomory method.

Solution: The mathematic model is assembled first. Let $x_1, x_2 \in \mathbb{Z}$ be the number of sold units of the first and second types, respectively. Then, the cash sales read

$$f(x_1, x_2) = 1000x_1 + 2000x_2$$

The function above returns a result in EUR, and it can easily be simplified to return a result in thousands of EUR (tEUR) as

$$f(x_1, x_2) = x_1 + 2x_2$$

The constraint on the overall steel material can be expressed as

$$-3x_1 + 4x_2 \leq 6$$

because the material quantity increases with selling the first-type product.

The plastic granules management reads

$$4x_1 + 3x_2 \leq 12$$

Let the task be solved without the integer condition first via the simple table. Note that the pivot elements in each step are highlighted in italics.

x_1	x_2	x_3	x_4	b_i
– 3	4	1	0	6
4	3	0	1	12
– 1	– 2	0	0	0

x_1	x_2	x_3	x_4	b_i
– 0.75	1	0.25	0	1.5
6.25	0	– 0.75	1	7.5
– 2.5	0	0.5	0	3

x_1	x_2	x_3	x_4	b_i
0	1	0.16	0.12	2.4
1	0	– 0.12	0.16	1.2
0	0	0.2	0.4	6

Hence, the optimal non-integer solution is

$$f\left(\mathbf{x}_R^*\right) = 6$$
$$x_1 = 1.2, x_2 = 2.4, x_3 = x_4 = 0$$

Follow Algorithm 4.5. Set $i = 1$, and find $j_{\max}$ as per (4.20). Apparently, it is x_2, for which one can write

$$x_2 = \lfloor x_2 \rfloor + \{x_2\} = 2 + 0.4$$

As the position of 1 in the 2nd column (i.e., below x_2) is also one, then $k_{\max} = 1$. Now look at numbers in the 1st row of the simplex table. The coefficients $a_{k_{\max},3}$ and $a_{k_{\max},4}$ under non-basic variables x_3 and x_4, respectively, are positive; hence, (4.21) yields

$$0.16x_3 + 0.12x_4 - x_{s,1} = 0.4$$

This condition is multiplied by – 1 and added to the simplex table (the added row and column are highlighted in bold)

x_1	x_2	x_3	x_4	$x_{s,1}$	b_i
0	1	0.16	0.12	**0**	2.4
1	0	– 0.12	0.16	**0**	1.2
0	**0**	**– 0.16**	**– 0.12**	**1**	**– 0.4**
0	0	0.2	0.4	**0**	6

Then a step of the dual method as per Algorithm 4.3 is performed as

x_1	x_2	x_3	x_4	$x_{s,1}$	b_i
0	1	0.16	0.12	0	2.4
1	0	– 0.12	0.16	0	1.2
0	0	*– 0.16*	– 0.12	1	– 0.4
0	0	0.2	0.4	0	6

The pivot element is highlighted. It is calculated and determined using (4.17) as

$$\max_{\leq 0}\{0/0,\ 0/0,\ 0.2/-0.16,\ 0.4/-0.12\} = 0.2/-0.16 = -5/4$$

Then, the pivot element is eliminated via step 6 of Algorithm 4.1, giving rise to

x_1	x_2	x_3	x_4	$x_{s,1}$	b_i
0	1	0	0	1	2
1	0	0	0.25	– 0.75	1.5
0	0	1	0.75	– 6.25	2.5
0	0	0	0.25	1.25	5.5

It this step, the deduced result reads:

$$f\left(\mathbf{x}^*_{R,1}\right) = 5.5$$
$$x_1 = 1.5, x_2 = 2, x_3 = 2.5, x_4 = x_{s,1} = 0$$

It is worth noting that the objective function value decreased. Generally, it cannot increase in any step of solving the ILP with a fitness function (i.e., a maximizing task). Now, it holds that $x_2 \in \mathbb{Z}$; nevertheless, x_1 is a non-integer.

Set $i = 2$. It can be determined from the table that $k_{\max} = 2$. Another condition (constraint) is introduced in accordance to (4.21):

$$0.25x_4 + 0.25x_{s,1} - x_{s,2} = 0.5$$

Notice that the coefficient below $x_{s,1}$ in the 2nd row is negative (– 0.75), therefore it is considered $0.25 = -0.75 + 1$ in the new constraint. Note that this operation is reasonable since

$$\underbrace{-0.75x_{s,1}+x_{s,1}}_{0.25x_{s,1}}\underbrace{-x_{s,1}-x'_{s,2}}_{-x_{s,2}}$$

In other words, $x_{s,2}$ remains an integer, but its value is increased by $x_{s,1}$ compared to the actual one (denoted by $x'_{s,2}$).

Hence, a new form of the simplex table is assembled and a step of the dual method is performed again:

x_1	x_2	x_3	x_4	$x_{s,1}$	$x_{s,2}$	b_i
0	1	0	0	1	**0**	2
1	0	0	0.25	– 0.75	**0**	1.5
0	0	1	0.75	– 6.25	**0**	2.5
0	**0**	**0**	*– 0.25*	**– 0.25**	**1**	**– 0.5**
0	0	0	0.25	1.25	**0**	5.5

x_1	x_2	x_3	x_4	$x_{s,1}$	$x_{s,2}$	b_i
0	1	0	0	1	0	2
1	0	0	0	– 1	1	1
0	0	1	0	– 7	3	1
0	0	0	1	1	– 4	2
0	0	0	0.25	1.25	0	5.5

The pivot column selection can be interpreted as a negative ratio that is "the closest to zero", as per (4.17).

The obtained solution (see step 7 of Algorithm 4.1) is integer. Namely:

$$f\left(\mathbf{x}_Z^*\right)=5$$
$$x_1=1, x_2=2, x_3=1, x_4=2, x_{s,1}=x_{s,2}=0$$

That is, the company produces one unit of a good for the metallurgy company and two units of another product, and the cash sales are tEUR 5 = EUR 5000. It can easily be verified that 1 q of steel material and 2 q of plastic granules are left in stock.

It is worth noting that $x_{s,1}=x_{s,2}=0$ means that $x_1=1$ and $x_2=2$ are simply obtained by removing the non-integer parts from the initial (non-integer) guess.

Example 4.8 *Task*: A hungry tiger escaped from the zoo. It only has 10 h to hunt (that is exactly until dawn—he has to spend the rest of the day in hiding). Its current detection rate corresponds to an index of 0. If its detection rate reaches an index of 6, it will be caught by the zookeeper.

Pheasants are easy and relatively quick prey for tigers. It can find and choke a hen in just 1 h. In addition, its detection rate increases by 1 point when running around the forest. Unfortunately, such a snack will not fill it up very much.

A sheep will satisfy a tiger 5× more than a hen pheasant, but it costs 4 h to find and digest it. In addition, there is a risk that the sheep will bleat loudly, increasing the detection rate by 2 points.

What prey composition should it choose to maximize its satiety? Will it be caught before dawn?

Solution: Let the feeling of satiety after consuming the pheasant be of weight 1, while that when eating a sheep be of weight 5. The hunting takes 1 and 4 h, respectively, and the limit is 10. The rates are 1 and 2, respectively, and the limit is 6. Hence, the ILP problem can be formulated as follows:

$$\max f(\mathbf{x}) = x_1 + 5x_2$$

s.t.

$$x_1 + 4x_2 \leq 10$$
$$x_1 + 2x_2 \leq 6$$

where $x_1, x_2 \in \mathbb{Z}$ are the numbers of hunted pheasants and sheep, respectively.

Algorithm 4.1 gives the following results (of the related non-integer LP task):

x_1	x_2	x_3	x_4	b_i
1	4	1	0	10
1	2	0	1	6
− 1	− 5	0	0	0

x_1	x_2	x_3	x_4	b_i
0.25	1	0.25	0	2.5
0.5	0	− 0.5	1	1
0.25	0	1.25	0	12.5

The non-integer LP has the optimal solution

$$f\left(\mathbf{x}_R^*\right) = 12.5$$
$$x_1 = 0, x_2 = 2.5, x_3 = 0, x_4 = 1$$

Naturally, the tiger cannot hunt down and kill "half of the sheep". Therefore, finding an optimal integer solution to the problem is necessary. Following Algorithm 4.5, the first row of the above-given table gives rise to the additional constraint

$$0.25x_1 + 0.25x_3 - x_{s,1} = 0.5$$

that is incorporated into the simplex table

x_1	x_2	x_3	x_4	$x_{s,1}$	b_i
0.25	1	0.25	0	**0**	2.5
0.5	0	– 0.5	1	**0**	1
– 0.25	**0**	**– 0.25**	**0**	**1**	**– 0.5**
0.25	0	1.25	0	**0**	12.5

x_1	x_2	x_3	x_4	$x_{s,1}$	b_i
0	1	0	0	1	2
0	0	– 1	1	2	0
1	0	1	0	– 4	2
0	0	1	0	0	12

Now, the ILP solution is already optimal with

$$f\left(\mathbf{x}_Z^*\right) = 12$$
$$x_1 = 2, x_2 = 2, x_3 = 0, x_4 = 0, x_{s,1} = 0$$

Hence, the tiger eats two hens and two sheep, which takes a full 10 h as slack variable x_4 is zero. Therefore, it is not caught before dawn.

Notice that, albeit the function value has not significantly changed, the optimal value of $x_1 \in \mathbb{Z}$ "jumped" from 0 to 2. This very simple example demonstrates how far the optimal integer solution can be from the optimal non-integer one.

Another interesting observation is that the basic (slack) variable x_4 is zero.

References

1. Chong, E.K.P., Lu, W.S., Zak, S.H.: An Introduction to Optimization, 5th edn. Wiley, New York (2023)
2. Dantzig, G.B.: Linear Programming and Extensions. Princeton, New Jersey (1998)
3. Fischetti, M.: Introduction to Mathematical Optimization. Independently Published (2019)
4. Fletcher, R.: Practical Methods of Optimization, 2nd edn. Wiley, New York (1987)
5. French, M.: Fundamentals of Optimization: Methods, Minimum Principles, and Applications for Making Things Better. Springer, Cham (2018)
6. Gass, S.I.: Linear Programming: Methods and Applications, 5th edn. McGraw-Hill Book Company, New York (1984)
7. Gill, P.E., Murray, W., Wright, M.H.: Practical Optimization. Academic Press, London (1981)
8. Guenin, B., Könemann, J., Tunçel, L.: A Gentle Introduction to Optimization. Cambridge University Press, Cambridge (2014)
9. Hillier, F.S., Lieberman, G.J.: Introduction to Operations Research, 7th edn. McGraw-Hill, New York (2002)

10. Lange, K.: Optimization, 2nd edn. Springer, New York (2013)
11. Luenberger, D.G., Ye, Y.: Linear and Nonlinear Programming, 3rd edn. Springer, New York (2008)
12. Rardin, R.L.: Optimization in Operations Research, 2nd edn. Pearson, London (2016)
13. Tabak, D., Kuo, B.C.: Optimal Control by Mathematical Programming. Prentice-Hall, New York (1971)
14. Vershik, A.: L. V. Kantorovich and linear programming. https://doi.org/10.48550/arXiv.0707.0491 (2007)
15. Walsh, G.R.: Methods of Optimization. Wiley, New York (1975)
16. Winston, W.L.: Operations Research: Applications and Algorithms. Brooks/Cole—Thomson Learning, Belmond (2004)
17. Wolsley, L.A.: Integer Programming. Wiley-Interscience, New York (1998)

Chapter 5
Dynamic Programming

A class of problems solved by dynamic programming (DP) methods [14, 15] represents another specific branch of optimization dealing with inequality constraints. The reign of DP covers a wide range of problems and techniques; however, the solution to the problems expressed by a generally non-linear (separable) objective function with a linear constraint in the form of inequalities will only be introduced in this chapter. Hence, it constitutes a more general task than the LP problem.

The solution using DP is based on decomposing a problem into subproblems, the solution of which is stored for further potential use. The method is particularly suitable for tasks that can be divided into subtasks that are similar and can be repeated. In this sense, DP belongs to the class of so-called separable programming. The solution takes two steps: *Direct calculating* and *backward deducing the results*.

Two problem formulations will be presented in this chapter. First is the so-called tabular form of DP, where the analytically specified problem is solved in discrete steps using a table. The task may practically represent the problem of dividing "resources" (e.g., investments) to maximize the positive or minimize the negative impact of this division. Then, the so-called network form DP also called the transport problem, which mathematically expresses the search for the shortest (or longest) path in the evaluated oriented graph, will be presented. Here, the use is evident. Readers are referred to [1, 3–12, 16, 17] for more details about DP in the context of other parameter optimization problems and techniques.

5.1 Bellman's Principle of Optimality and Particular DP Problem Formulation

Algorithms solving the DP optimization problems are based on the so-called **Bellman's principle of optimality** that can be formulated in several ways. The following Theorem 5.1 rephrases the principle to be understandable for practitioners.

L. Pekař, *Optimization: An Introduction*, Studies in Systems, Decision and Control 239,
https://doi.org/10.1007/978-3-031-86326-4_5

Theorem 5.1 (Bellman's principle of optimality)

1. *Regardless of whether the initial estimate and solution are made, the remaining solution constitutes an optimal solution concerning the state yielding from the first solution estimation.*
2. *The optimal strategy from any solution state to the solution of the entire problem does not depend on the way how the solution got to this current point.*
3. *A sub-strategy of the optimal strategy is also optimal.*

Bellman's principle yields the following facts. An optimal solution to the DP problem can be obtained by successive appending an optimal solution of the problem subtasks to the currently existing "encapsulated" partial solution. This computational stage can be seen as a direct pass through the solution. Once the entire problem is solved, the backward pass enables the reading of the results.

From a wide range of DP problems, problems with a generally nonlinear separable objective function with a linear constraint are concisely presented in this chapter, i.e.:

$$\operatorname{extr} f(\mathbf{x}) = f_1(x_1) + f_2(x_2) + \cdots + f_n(x_n) = \sum_{i=1}^{n} f_i(x_i) \in \mathbb{R} \tag{5.1}$$

s.t.:

$$\begin{aligned} &\sum_{i=1}^{n} a_i x_i \leq b \\ &a_i, x_i \geq 0, i = 1, 2, \ldots, n \end{aligned} \tag{5.2}$$

Hence, task (5.1)–(5.2) represents a specific multi-dimensional inequality constrained problem, see Sect. 2.4.

5.2 The Tabular-Form Method

Consider problem (5.1)–(5.2) with $a_i = 1, i = 1, 2, \ldots, n$, for simplicity. The tabular-form method represents a combined analytic-numerical technique that uses a table to solve the given DP problem in discrete steps. The obtained result is not an analytically exact solution, but it constitutes an approximate estimate given by the discretization of the right-hand side b of constraint (5.2):

$$\begin{aligned} &0 = b_0 < b_1 < b_2 < \ldots < b_m = b \\ &h_j := b_{j+1} - b_j, j = 0, 1, \ldots, m-1 \end{aligned} \tag{5.3}$$

Hence, a necessary condition is that discretization (5.3) exists (namely, it is given by the particular problem assignment). Note that the discretization step h_j in (5.3) can vary; however, it is mostly assumed that $h_j = h =$ const. in this section (for simplicity).

An algorithm sketch followed by a Matlab code for a minimization problem is given to readers.

Algorithm 5.1 (Tabular-form DP method)

1. The "source" b is discretized via (5.3).
2. Set

$$\begin{aligned} F_1(b_j) &:= f_1(x_{1,j}),\, x_{1,j} = b_j \\ x^*_{1,j} &:= x_{1,j},\ \ j = 0, 1, \ldots, m \end{aligned} \tag{5.4}$$

3. Initiate the iteration counter $i = 2$. For each "resource discrete value" b_j (concerning the discretization step h_j), its optimal distribution between variables $x_1, x_2, \ldots, x_{i-1}$ and variable x_i is made according to this policy:

   (a) If the constraint of the $\leq$ type is in (5.2), then

$$\begin{aligned} F_i(b_j) &:= \underset{0 \leq x_i \leq b_j}{\text{extr}} \left\{ f_i(x_i) + F_{i-1}(b_j - x_i) \right\} \\ x^*_{i,j} &:= \underset{x_i}{\arg} F_i(b_j) \\ j &= 0, 1, \ldots, m \end{aligned} \tag{5.5}$$

   (b) If the constraint of the $=$ type is in (5.2), then

$$\begin{aligned} & F_i(b_j) := \underset{0 \leq x_i \leq b_j}{\text{extr}} \left\{ f_i(x_i) + F_{i-1}(b_j - x_i) \right\} \\ & x^*_{i,j} = \underset{x_i}{\arg} F_i(b_j) \\ & \text{if } i = 2, \ldots, n-1 \text{ then } \ j = 0, 1, \ldots, m \\ & \text{if } i = n \qquad\qquad\ \text{then } \ j = m \end{aligned} \tag{5.6}$$

   The results are inserted into Table 5.1.
4. If $i < n$, set $i = i + 1$; else (i.e., $i = n$), go to step 5.
5. Results deduction:

   (a) The optimal function value:

$$f^* = \underset{j=0,1,\ldots,m}{\text{extr}} F_n(b_j) \tag{5.7}$$

   (i.e., the extremal value in the right-most column)

Table 5.1 The tabular-form DP method

b_j	$x_1(b_j)$	$F_1(b_j)$	$x_2(b_j)$	$F_2(b_j)$	...	$x_n(b_j)$	$F_n(b_j)$
$b_0 = 0$	$x^*_{1,0}$	$F_1(b_0)$	$x^*_{2,0}$	$F_2(b_0)$	...	$x^*_{n,0}$	$F_n(b_0)$
b_1	$x^*_{1,1}$	$F_1(b_1)$	$x^*_{2,1}$	$F_2(b_1)$	...	$x^*_{n,1}$	$F_n(b_1)$
$\vdots$	$\vdots$	$\vdots$	$\vdots$	$\vdots$	$\ddots$	$\vdots$	$\vdots$
b_{m-1}	$x^*_{1,m-1}$	$F_1(b_{m-1})$	$x^*_{2,m-1}$	$F_2(b_{m-1})$	...	$x^*_{n,m-1}$	$F_{n,}(b_{m-1})$
$b_m = b$	$x^*_{1,m}$	$F_1(b_m)$	$x^*_{2,m}$	$F_2(b_m)$	...	$x^*_{n,m}$	$F_n(b_m)$

(b) Optimal $\mathbf{x}^*$:

Define $j^* = \arg\text{extr}_{j=0,1,\ldots,m} F_n(b_j)$ (i.e., the row number with the optimal objective function value). Let $b_{j^*,n} = b_{j^*}$, $j^*_n = j^*$, Then, the results are deduced by using the successive backward steps:

$$x^*_n = x^*_{n,j^*_n} \tag{5.8}$$

$$\begin{aligned} b_{j^*,n-1} &= b_{j^*,n} - x^*_{n,j^*_n} \\ j^*_{n-1} &= \arg b_{j^*,n-1} \end{aligned} \tag{5.9}$$

(i.e., j^*_{n-1} is the row with $b_{j^*,n-1}$ in the left-most column)

$$x^*_{n-1} = x^*_{n-1,j^*_{n-1}} \tag{5.10}$$

$$\begin{aligned} b_{j^*,n-2} &= b_{j^*,n-1} - x^*_{n-1,j^*} \\ j^*_{n-2} &= \arg b_{j^*,n-2} \end{aligned} \tag{5.11}$$

$$x^*_{n-2} = x^*_{n-2,j^*_{n-2}} \tag{5.12}$$

Etc., until x^*_1.

Remarks on Algorithm 5.1:

- Step 5 of the algorithm goes back through the columns in Table 5.1. The particular value of the optimal variable is subtracted from the current value of the discretized resource. The result (i.e., the "remainder of the resource") determines the row in which the value of the next variable (i.e., with the index decreased by 1) is determined, etc.
- In the equality ("=") case of (5.2), the right-most double-column includes only the last row (see step 3 of Algorithm 5.1). That is, the full resource is eventually distributed among the variables. The particular values in the left and right-hand sub-columns mean x^*_n and $f(\mathbf{x}^*)$, respectively. Contrariwise, for the inequality

("<") case, it is necessary to fill in all the rows of the right-most double-column as it is not a priori known which part of the resource will eventually be distributed.
- This DP problem always has a solution. However, it is not necessarily unique; see (5.5) or (5.6).
- This method can, e.g., solve the task of the "distribution of investments" type.
- The application of the Bellman principle (Theorem 5.1) is obvious. The feedforward computation is composed of a successive winding of the solution up. Once Table 5.1 is filled in, the optimal results are deduced by a reverse reading. Note that partial optimal sub-results are given by the values of $F_i(b_j)$.
- A solution procedure to the tasks with a weighted constraint (i.e., $\exists i : a_i \neq 1$) is unambiguous. Suppose the task assignment is not supplemented with a verbal statement of the problem. In that case, it is not clear whether the division h_j of the source is related to individual variables x_i or the entire "weight" (or "sub-resource") for a particular variable in the constraint, i.e., $a_i x_i$.

In the former case, some resource discretizations might not be available or allowable. For instance, if the constraint reads

$$2x_1 + 3x_2 = 4, \ h = 1 \tag{5.13}$$

where h represents the discretization of x_1, x_2, then already the distribution of $b_1 = 1$ is impossible, because

$$2 \cdot 1 + 3 \cdot 0 > 1$$

In the latter case, a different resolution for a different variable can be obtained. For instance, considering constraint (5.13), where h represents a possible resource distribution between $2x_1$ and $3x_2$ now, the resolution of x_1 is 0.5, while that of x_2 is 1/3.

```
% INPUTS
n = input('Enter the number of variables (n): ');

syms x real
fs = strings(1,n);
for i = 1:1:n
    hint = sprintf('Enter the sub-function number %d in x (e.g.,
x^2+1):', i);
   fs(i) = input(hint);
end

b = input('Enter the source value (b): ');
h = input('Enter the discretization step (h): ');
inq = input('Enter the (in)equality type of the
constraint (1 = "=", 2 = "<="): ');

% INIT
row_num = round(b/h+1);
```

```
col_num = 2*n+1;
DPTable = zeros(row_num,col_num);

for j = 1:1:row_num % THE RESOURCE DISCRETIZATION AND THE LEFT-MOST
DOUBLE-COLUMN
   f = fs(1); f = eval(f);
   DPTable(j,1:3) = [(j-1)*h, (j-1)*h, subs(f,x,(j-1)*h)];
end

% DOUBLE-COLUMNS EXCEPT FOR THE RIGHT-MOST ONE
for i = 4:2:(col_num-3)
   f = fs(i/2); f = eval(f);
   for j = 1:1:row_num % bj = (j-1)*h
      f_min = inf; x_min = 0;
      for k = 1:1:j
         X = subs(f,x,(k-1)*h)+DPTable(j-k+1,i-1);
         if X<f_min
            f_min = X;
            x_min = (k-1)*h;
         end
      end
      DPTable(j, i:i+1) = [x_min,f_min]; % F(i/2,j)
   end

end

% THE RIGHT-MOST DOUBLE-COLUMN
f = fs(n); f = eval(f);
f_min = inf; x_min = 0;
for k = 1:1:row_num
   X = subs(f,x,(k-1)*h)+DPTable(row_num-k+1,col_num-2);
   if X<f_min
      f_min = X;
      x_min = (k-1)*h;
   end
end
DPTable(row_num, col_num-1:col_num) = [x_min,f_min];
if inq==2 % THE "<=" TYPE
   for j = 1:1:(row_num-1)
      f_min = inf; x_min = 0;
      for k = 1:1:j
         X = subs(f,x,(k-1)*h)+DPTable(j-k+1,col_num-2);
         if X<f_min
            f_min = X;
            x_min = (k-1)*h;
         end
      end
      DPTable(j, col_num-1:col_num) = [x_min,f_min];
   end
end

% RESULTS DEDUCTION - THE BACKWARD PASS
x_opt = zeros(n,1);
if inq==1
```

```
    x_opt(n) = DPTable(row_num, col_num-1); f_opt = DPTable(row_num,
col_num);
    j_opt = row_num;
else
    [f_opt, j_opt] = min(DPTable(1:row_num, col_num));
    x_opt(n) = DPTable(j_opt, col_num-1);
end
for i=(n-1):-1:1
    bi_opt = DPTable(j_opt, 1) - x_opt(i+1);
    j_opt = bi_opt/h+1;
    x_opt(i) = DPTable(j_opt, 2*i);
end

% OUTPUTS
% OPTIMAL COST FUNCTION VALUE
str = sprintf("The optimal f = %0.5g", f_opt);
disp(str);
disp('The optimal variable values are:');
for i = 1:1:n
    str = sprintf('x%d = %0.5g', i, x_opt(i));
    disp(str);
end
```

A real-life-inspired numerical example follows.

Example 5.1 *Task*: Johnny is a multimillionaire and owns a swimming pool in his modest villa. Every day, he has to throw in 5 tablets with chemicals that, in contact with water, develop chlorine in gaseous form. As Johnny has children and pets, he wants to pollute his house and surroundings as little as possible. Temperature and air pressure have the most significant influence on release. Johnny deduced that the smell was different at different times of the day and had his clever chemical physicist friend Mary analyze the volume of chlorine released. After many trials and regression analysis, Mary found that in the morning, the relationship between the amount of chemical and the released chlorine gas is expressed by the approximate relationship $0.5x^2$ [m^3], where x represents the number of tablets. At noon, the addiction was only $0.25x^2$, and in the evening, it was $0.1x^3$. Hence, Mary was given the task by Johnny to figure out how many tablets to throw in at what time so that the total daily volume of gas was as small as possible. How did clever Mary solve it?

Solution: A mathematical formulation of the problem ought to be assembled first. Let x_1 be the number of tablets used in the morning, and x_2 and x_3 be the number of tablets thrown in at noon and in the evening, respectively. Then, the model reads:

$$\min f(x_1, x_2, x_3) = 0.5x_1^2 + 0.25x_2^2 + 0.1x_3^3 = f_1(x_1) + f_2(x_2) + f_3(x_3)$$

s.t.:

$$x_1 + x_2 + x_3 = 5$$

Follow the tabular method as in Algorithm 5.1. Apparently, $b = 5$, $h = 1$. Consider x_1 and the corresponding sub-function $f_1(x_1)$ first (i.e., $i = 1$). Now, each discretized part of b ($b_i = 0, 1, 2, 3, 4, 5$) is devoted solely for:

$$F_1(b_j) := f_1(x_{1,j}), x_{1,j} = b_j$$
$$x^*_{1,j} := x_{1,j}, \quad j = 0, 1, \ldots, 5$$

Namely, if $b_0 = 0$, then $F_1(b_0) = f_1(x_{1,0}) = f_1(0) = 0$, and inevitably $x^*_{1,0} = 0$. For $b_1 = 1$, one has $F_1(b_1) = f_1(x_{1,1}) = f_1(1) = 0.5$, and $x^*_{1,1} = 1$. For $b_2 = 2$, it is $F_1(b_2) = f_1(x_{1,2}) = f_1(2) = 2$ and $x^*_{1,2} = 2$, etc. These results are inserted in the main table:

b_j	$x_1(b_j)$	$F_1(b_j)$	$x_2(b_j)$	$F_2(b_j)$	$x_3(b_j)$	$F_3(b_j)$
0	0	0				
1	1	0.5				
2	2	2				
3	3	4.5				
4	4	8				
5	5	12.5				

Set $i = 2$ and distribute particular b_j optimally between $F_1(b_j)$ and $f_2(x_2)$, i.e., x_1 and x_2, respectively. $F_1(b_j)$ represents a set of the "fixed best results so far". Mathematically expressed:

$$F_2(b_j) := \min_{0 \le x_2 \le b_j} \{f_2(x_2) + F_1(b_j - x_2)\}$$
$$x^*_{2,j} := \arg_{x_2} F_2(b_j)$$
$$j = 0, 1, \ldots, 5$$

In detail, let $b_0 = 0$, then it obviously holds that

$$F_2(b_0) = \min\{f_2(0) + F_1(0 - 0)\} = 0$$
$$x^*_{2,0} = 0$$

For $b_1 = 1$, the corresponding subtask reads:

$$F_2(b_1) = \min\{f_2(0) + F_1(1 - 0), f_2(1) + F_1(1 - 1)\}$$
$$= \min\{0 + 0.5, 0.25 + 0\} = 0.25$$
$$x^*_{2,1} = 1$$

This decision problem can be expressed by the following auxiliary table:

x_2	x_1	$f_2(x_2)$	$F_1(1-x_2)$	$F_2(1)$
0	1	0	0.5	0.5
1	*0*	*0.25*	*0*	*0.25*

The (partial) optimal solution is highlighted in italics. Notice that the 4th column can already be found in the overall partially filled table (i.e., from the optimal resource distribution for x_1).

The discretization of $b_2 = 2$ yields

x_2	x_1	$f_2(x_2)$	$F_1(2-x_2)$	$F_2(2)$
0	2	0	2	2
1	*1*	*0.25*	*0.5*	*0.75*
2	0	1	0	1

i.e., $x_{2,2}^* = 1$.

Now, it is to distribute $b_3 = 3$ between x_1 and x_2:

x_2	x_1	$f_2(x_2)$	$F_1(3-x_2)$	$F_2(3)$
0	3	0	4.5	4.5
1	2	0.25	2	2.25
2	*1*	*1*	*0.5*	*1.5*
3	0	2.25	0	2.25

which implies $x_{2,3}^* = 2$.

The remaining distributions of $b_4 = 4$ and $b_5 = 5$ are as follows:

x_2	x_1	$f_2(x_2)$	$F_1(4-x_2)$	$F_2(4)$
0	4	0	8	8
1	3	0.25	4.5	4.75
2	2	1	2	3
3	*1*	*2.25*	*0.5*	*2.75*
4	0	4	0	4

x_2	x_1	$f_2(x_2)$	$F_1(5-x_2)$	$F_2(5)$
0	5	0	12.5	12.5
1	4	0.25	8	8.25

(continued)

(continued)

x_2	x_1	$f_2(x_2)$	$F_1(5-x_2)$	$F_2(5)$
2	3	1	4.5	5.5
3	*2*	*2.25*	*2*	*4.25*
4	1	4	0.5	4.5
5	0	6.25	0	6.25

The results are added to the main table:

b_j	$x_1(b_j)$	$F_1(b_j)$	$x_2(b_j)$	$F_2(b_j)$	$x_3(b_j)$	$F_3(b_j)$
0	0	0	0	0		
1	1	0.5	1	0.25		
2	2	2	1	0.75		
3	3	4.5	2	1.5		
4	4	8	3	2.75		
5	5	12.5	3	4.25		

Increment the variable-passing counter to $i = 3$. As the given constraint is of the equality type, all three variables are fed by the "whole resource" b, not only by its part (which would be the case of an inequality-type constraint). Hence, it is necessary and sufficient to fill in the last row of the right-most double-column, i.e., to analyze the optimal distribution of $b = b_5 = 5$ between $x_1 + x_2$ (that has already been "encapsulated") and x_3 (that represents the newly added variable):

$$F_3(b_5) := \min_{0 \le x_3 \le b_5} \{f_3(x_3) + F_2(b_5 - x_3)\}$$

$$x^*_{3,5} := \arg_{x_3} F_3(b_5)$$

x_3	$x_1 + x_2$	$f_3(x_3)$	$F_2(5-x_3)$	$F_3(5)$
0	5	0	4.25	4.25
1	4	0.1	2.75	2.85
2	*3*	*0.8*	*1.5*	*2.3*
3	2	2.7	0.75	3.45
4	1	6.4	0.25	6.65
5	0	12.5	0	12.5

Then, the final main table is as follows:

b_j	$x_1(b_j)$	$F_1(b_j)$	$x_2(b_j)$	$F_2(b_j)$	$x_3(b_j)$	$F_3(b_j)$
0	0	0	0	0		
1	**1**	**0.5**	1	0.25		
2	2	2	1	0.75		
3	3	4.5	**2**	**1.5**		
4	4	8	3	2.75		
5	5	12.5	3	4.25	**2**	**2.3**

Now, the deduction of results is performed as per step 5 of Algorithm 5.1. For the given task, the optimal cost function value can be unambiguously found in the last row (as there is no other option here); i.e., $b_5 = 5$, $j^* = j_3^* = 5$, $x_3^* = x_{3,5}^* = 2$, and $f^* = 2.3$. This partial result is subtracted from the "resource": $b_5 - x_3^* = 5 - 2 = 3$, i.e., $j_2^* = 3 = b_{3,2} = b_3$. Hence, the number 3 must be distributed between x_1, x_2. In the 3rd row, the value $x_2^* = x_{2,3}^* = 2$ is found. This value is subtracted from the remaining resource again: $b_3 - x_2^* = 3 - 2 = 1$. Finally, it is clear that $x_1^* = 1$, which can be verified in the 1st row: $x_{1,1}^* = b_1 = 1$. Note that the above-described backward results reading is highlighted in bold.

Checking the obtained results:

$$f = 0.5x_1^{*2} + 0.25x_2^{*2} + 0.1x_3^{*3} = 2.3$$
$$x_1^{*2} + x_2^{*2} + x_3^{*3} = 1 + 2 + 2 = 5$$

What does the optimal solution mean for the practice? Johnny must throw in one tablet in the morning, two tablets at noon, and two tablets in the evening. The amount of 2.3 m^3 of chlorine gas is released into the air.

Notice that the number of calculations in the table is lower than all possible combinations of dividing the "resource" between individual variables. While each double box in the main table represents one new calculation (see sub-calculations in the auxiliary tables), of which there are a total of 13, you can easily verify that all possible divisions of the number 5 into three additions with the discrete step 1 are a total of 21. Naturally, the mentioned algorithm is asymptotically faster; the acceleration of the calculation is more pronounced with the increasing number of variables and the division of the "resource".

5.3 Searching for the Shortest Path—Dijkstra's Algorithm

This problem represents finding the shortest (or longest) path between two nodes (vertices) in a (non-negatively) weighted directed connected graph. It is an everyday task for most people to travel from point A to point B with minimum time or kilometers. The solution is provided by Dijkstra's algorithm, the author of which was the Dutch computer scientist Edsger Wybe Dijkstra (1930–2002), who published the algorithm in 1959 [2, 13].

First, the concept of a graph is concisely introduced. Then, the task of finding the shortest path and Dijkstra's algorithm are presented.

5.3.1 *Basic Notions*

Definition 5.1 A (simple) **graph** is a pair $G = \langle V, E \rangle$ where V is a set of *vertices* (*nodes*) and E represents a set of *edges* (arcs).

Definition 5.2 A **directed graph** has ordered pairs of distinct vertices, i.e., $E \subseteq V \times V$. It means that the graph contains edges between vertices with a certain direction (from-to).

Definition 5.3 In a **weighted graph**, each vertex is assigned some (usually real) number. If this number is non-negative, it is a *non-negatively* weighted graph.

Definition 5.4 A **path** of a graph is a sequence of joint vertices for which each vertex is visited only once.

Definition 5.5 In a **connected graph**, there exists a path between each pair of vertices.

A graph can be represented, e.g., by its drawing, which is suitable for didactic purposes (see Fig. 5.1, where a non-negatively weighted directed connected graph is displayed).

As an alternative, a graph can be represented by the so-called *adjacency matrix* **A** where value (entry) $a_{ij} \in \mathbf{A}$ represents the edge length between vertices $[i, j]$ where $\{i, j\} \in V$. If a particular edge does not exist or it the inverse direction, it is possible to set $a_{ij} = \infty$.

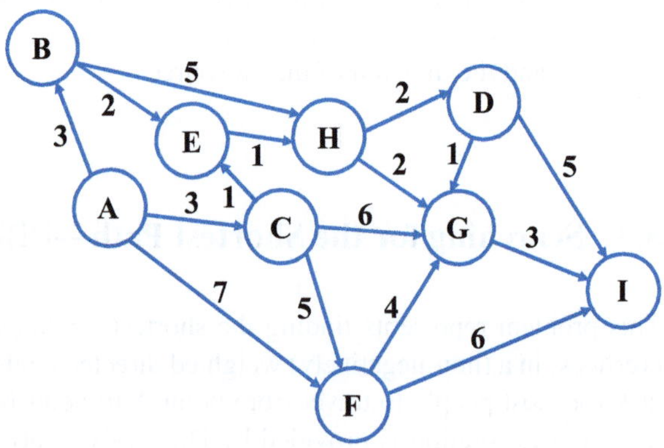

Fig. 5.1 An example of connected and non-negatively weighted directed graph

5.3.2 *Problem Formulation and the Algorithm*

As mentioned above, the task is to search for the shortest path between two vertices in a **non-negatively weighted directed connected graph**. The problem can be algebraically formulated as

$$\min \sum_{\{i,j\}\in V} a_{ij} = a_{sj_1} + a_{j_1,i_2} + a_{i_2,j_2} + a_{j_2,i_3} + \cdots + a_{i_n,j_n} + a_{j_n,f} \tag{5.14}$$

where $s \in V$ and $f \in V$ are the start and finish vertices, respectively. Recall that $a_{ij} \geq 0 \in E$ is the edge length between nodes i and j (i.e., in the direction from i to j).

The Dijkstra's algorithm solving problem (5.14) follows.

Algorithm 5.2 (Dijkstra's algorithm)

1. Let the graph be represented, e.g., by adjacency matrix $\mathbf{A}$ of the dimension $n \times n$ (where $n = |V|$ is the number of vertices) with its entries $a_{ij} \in \mathbb{R}$, the starting vertex $s \in V$ and the final one $f \in V$.

   Introduced two sets of vertices: A set of *undiscovered* nodes, U, and a set of *discovered* ("encapsulated") nodes, D. Elements of the set represent vertex indices.

   In addition, let a vector $\mathbf{p} = \left[p_1, p_2, \ldots, p_n\right]^T \in \mathbb{R}^n$ of *predecessors* in the solution path be introduced. In the vector, $p_i \subset \mathbf{p}$ means the index of the predecessor of the ith vertex in the path. Finally, path lengths to particular vertices are stored in vector $\mathbf{d} = [d_1, d_2, \ldots, d_n]^T \in \mathbb{N}^n$ (i.e., $d_i \in \mathbf{d}$ expresses the current distance of the ith vertex from the starting one).

   Without loss of generality, let $s = 1$ and $f = n$.
2. Data sets initialization:

$$\begin{aligned} U &= V \\ D &= \emptyset \\ \mathbf{p} &= [\,]^T \\ \mathbf{d} &= [0, \infty, \infty, \ldots, \infty]^T \end{aligned} \tag{5.15}$$

3. Find a vertex $i_{\min}$ of U that has a minimum distance from the starting node:

$$i_{\min} = \arg\min_{i \in U} d_i \in \mathbf{d} \tag{5.16}$$

   and this vertex is added to D and taken out from U:

$$\begin{aligned} D &= D \cup i_{\min} \\ U &= U \backslash i_{\min} \end{aligned} \tag{5.17}$$

4. If $i_{\min} = f = n$, then go to step 6; else, go to step 5.
5. Find all vertices j directly connected to $i_{\min}$, i.e., those edges for which it holds that

$$a_{i_{\min},j} < \infty \tag{5.18}$$

For all these vertices, it is computed that

$$\overline{d}_j = d_{i_{\min}} + a_{i_{\min},j} \tag{5.19}$$

If $\overline{d}_j < d_j$, set $d_j = \overline{d}_j$ and $p_j = i_{\min}$.
Go to step 3.

6. Results deduction (the backward pass): The value of d_n equals the minimum path length. This minimum path can be deduced by following the predecessors from the final vertex to the starting one.

$$j_1 = p_n \rightarrow j_2 = p_{j_1} \rightarrow j_3 = p_{j_2} \rightarrow \ldots \rightarrow j_{m-1} = p_{j_{m-2}} \rightarrow j_m = s = 1 = p_{j_{m-1}} \tag{5.20}$$

Note that if condition $\overline{d}_j \leq d_j$ is considered in step 5 of Algorithm 5.2, it is possible to assign a set (i.e., more than one) of predecessors p_j to each node, which may yield finding all the minimum paths of the same length.

It is obvious that Dijkstra's algorithm agrees with Bellman's principle of successive "packaging" and "encapsulation" of partial optimal solutions from the starting vertex to the final one, followed by the backward determining the optimal path.

A Matlab code for Algorithm 5.2 follows.

```
% INPUT
A = input('Enter the adjacency matrix (A): ');

% INIT
n = length(A);
U = 1:1:n;
D = zeros(1,n);
p = zeros(1,n);
d = [0, inf*ones(1,n-1)];

% FORWARD PASS
while (D(n)==inf)
   d_min = inf;
   for i=1:1:n
      if (U(i)~=0) && (d(i)<d_min)
         d_min = d(i);
         V_ind = i;
      end
   end
   for j=1:1:n
      if (A(V_ind,j)<inf) && (j~=V_ind)
```

```
            d_ = d_min+A(V_ind,j);
            if d_<d(j)
               d(j) = d_;
               p(j) = V_ind;
            end
         end
      end
      D(V_ind) = V_ind;
      U(V_ind) = zeros;
   end

   % BACKWARD PASS AND OUTPUTS
   path_ = n;
   V_path = p(end);
   while V_path~=1
      path_ = [V_path path_];
      V_path = p(V_path);
   end
   path_ = [1 path_];
   str = sprintf("The minimum path length is %0.5g", d(end));
   disp(str);
   str = "";
   disp('The minimum path vertices are:');
   for i = 1:1:(length(path_)-1)
      str = str+sprintf('%d -> ', path_(i));
   end
   str = str+sprintf('%d', path_(end));
   disp(str),
```

Example 5.2 *Task*: Find the shortest path between nodes A and I in the graph displayed in Fig. 5.1 using Dijkstra's algorithm.

Solution: For simplicity and consistency with Algorithm 5.2, add integer indexes to the vertices instead of capital letters, i.e., use the mapping

$$\{A, B, C, D, E, F, G, H, I\} \rightarrow \{1, 2, 3, 4, 5, 6, 7, 8, 9\}$$

The adjacency matrix corresponding to the displayed graph has the form

$$A = \begin{bmatrix} 0 & 3 & 3 & \infty & \infty & 7 & \infty & \infty & \infty \\ \infty & 0 & \infty & \infty & 2 & \infty & \infty & 5 & \infty \\ \infty & \infty & 0 & \infty & 1 & 5 & 6 & \infty & \infty \\ \infty & \infty & \infty & 0 & \infty & \infty & 1 & \infty & 5 \\ \infty & \infty & \infty & \infty & 0 & \infty & \infty & 1 & \infty \\ \infty & \infty & \infty & \infty & \infty & 0 & 4 & \infty & 6 \\ \infty & \infty & \infty & \infty & \infty & \infty & 0 & \infty & 3 \\ \infty & \infty & \infty & 2 & \infty & \infty & 2 & 0 & \infty \\ \infty & \infty & \infty & \infty & \infty & \infty & \infty & \infty & 0 \end{bmatrix}$$

Perform the initialization first. The sets of undiscovered and discovered nodes are

$$U = V = \{1, 2, 3, 4, 5, 6, 7, 8, 9\}$$
$$D = \emptyset$$

respectively. The initial vectors of distances (partial path lengths) and predecessors read

$$\mathbf{d} = [0, \infty, \infty, \infty, \infty, \infty, \infty, \infty, \infty]^T$$
$$\mathbf{p} = [\,]^T$$

respectively.

As per step 3 of the algorithm, find the vertex in U with a minimum distance. It is apparently the starting one, i.e., $i_{\min} = 1$. Move this node to the set of discovered ones:

$$D = D \cup \{1\} = \{1\}$$
$$U = U \backslash \{1\} = \{2, 3, 4, 5, 6, 7, 8, 9\}$$

Now, it is to find all the direct edges from this vertex to its neighbors, i.e., edges governed by the adjacency matrix entries a_{12}, a_{13}, a_{16}. Let the path distance to the corresponding neighbors 1, 3, 6 be updated as follows:

$$\overline{d}_2 = d_1 + a_{12} = 0 + 3 = 3 < \infty = d_2 \Rightarrow d_2 = 3$$
$$\overline{d}_3 = d_1 + a_{13} = 0 + 3 = 3 < \infty = d_3 \Rightarrow d_3 = 3$$
$$\overline{d}_6 = d_1 + a_{16} = 0 + 7 = 7 < \infty = d_6 \Rightarrow d_6 = 7$$

$$\mathbf{d} = [0, 3, 3, \infty, \infty, 7, \infty, \infty, \infty]^T$$

The updated vector of predecessors reads

$$\mathbf{p} = [\emptyset, 1, 1, \emptyset, \emptyset, 1, \emptyset, \emptyset, \emptyset]^T$$

where $\emptyset$ means an empty element (e.g., the NULL pointer). Now, the first step of Algorithm 5.2 is finished. The selection of the undiscovered vertex with the minimum path length follows again. There are two options here since $.d_2 = d_3 = 3$. Without loss of generality, let $i_{\min} = 2$, and move the particular vertex to the discovered ones:

$$D = D \cup \{2\} = \{1, 2\}$$
$$U = U \backslash \{2\} = \{3, 4, 5, 6, 7, 8, 9\}$$

All the one-step paths from vertex 2 are found in $\mathbf{A}$: a_{25}, a_{28}. Update the corresponding distances from the starting vertex and the vector of predecessors:

$$\overline{d}_5 = d_2 + a_{25} = 3 + 2 = 5 < \infty = d_5 \Rightarrow d_5 = 5$$
$$\overline{d}_8 = d_2 + a_{28} = 3 + 5 = 8 < \infty = d_8 \Rightarrow d_8 = 8$$
$$\mathbf{d} = [0, 3, 3, \infty, 5, 7, \infty, 8, \infty]^T$$
$$\mathbf{p} = [\emptyset, 1, 1, \emptyset, 2, 1, \emptyset, 2, \emptyset]^T$$

Let another step be shown in detail. As $d_3 = \min_{\arg d \in U} \mathbf{d} = 3$, one gets

$$D = D \cup \{3\} = \{1, 2, 3\}$$
$$U = U \backslash \{3\} = \{4, 5, 6, 7, 8, 9\}$$

Distance from $i_{\min} = 3$ to its neighbors: $a_{35} = 1, a_{36} = 5, a_{37} = 6$. Hence

$$\overline{d}_5 = d_3 + a_{35} = 3 + 1 = 4 < 5 = d_5 \Rightarrow d_5 = 4$$
$$\overline{d}_6 = d_3 + a_{36} = 3 + 5 = 8 > 7 = d_6$$
$$\overline{d}_7 = d_3 + a_{37} = 3 + 6 = 9 < \infty = d_7 \Rightarrow d_7 = 9$$
$$\mathbf{d} = [0, 3, 3, \infty, 4, 7, 9, 8, \infty]^T$$
$$\mathbf{p} = [\emptyset, 1, 1, \emptyset, 3, 1, 7, 2, \emptyset]^T$$

Notice that the current distance and predecessor for vertex 6 are not updated.

The remaining iteration steps repeating steps 3–5 of Algorithm 5.2 are summarized in the following table:

Iteration step	$\min_{\arg d \in U} \mathbf{d}$	D, U	$\overline{d}.$	$\mathbf{d}, \mathbf{p}$
4	$d_5 = 4$	$D = \{1, 2, 3, 5\}$ $U = \{4, 6, 7, 8, 9\}$	$\overline{d}_8 = 5$	$\mathbf{d} = [0, 3, 3, \infty, 4, 7, 9, 5, \infty]^T$, $\mathbf{p} = [\emptyset, 1, 1, \emptyset, 3, 1, 3, 5, \emptyset]^T$
5	$d_8 = 5$	$D = \{1, 2, 3, 5, 8\}$ $U = \{4, 6, 7, 9\}$	$\overline{d}_4 = 7$ $\overline{d}_7 = 7$	$\mathbf{d} = [0, 3, 3, 7, 4, 7, 7, 5, \infty]^T$, $\mathbf{p} = [\emptyset, 1, 1, 8, 3, 1, 8, 5, \emptyset]^T$
6	$d_4 = 7$ (or possibly d_6, d_7)	$D = \{1, 2, 3, 4, 5, 8\}$ $U = \{6, 7, 9\}$	$\overline{d}_9 = 12$	$\mathbf{d} = [0, 3, 3, 7, 4, 7, 7, 5, 12]^T$, $\mathbf{p} = [\emptyset, 1, 1, 8, 3, 1, 8, 5, 4]^T$
7	$d_6 = 7$ (or possibly d_8)	$D = \{1, 2, 3, 4, 5, 6, 8\}$ $U = \{7, 9\}$	$\overline{d}_7 = 11$ $\overline{d}_9 = 13$	$\mathbf{d} = [0, 3, 3, 7, 4, 7, 7, 5, 12]^T$, $\mathbf{p} = [\emptyset, 1, 1, 8, 3, 1, 8, 5, 4]^T$
8	$d_7 = 7$	$D = \{1, 2, 3, 4, 5, 6, 7, 8\}$ $U = \{9\}$	$\overline{d}_9 = 10$	$\mathbf{d} = [0, 3, 3, 7, 4, 7, 7, 5, 10]^T$, $\mathbf{p} = [\emptyset, 1, 1, 8, 3, 1, 8, 5, 7]^T$

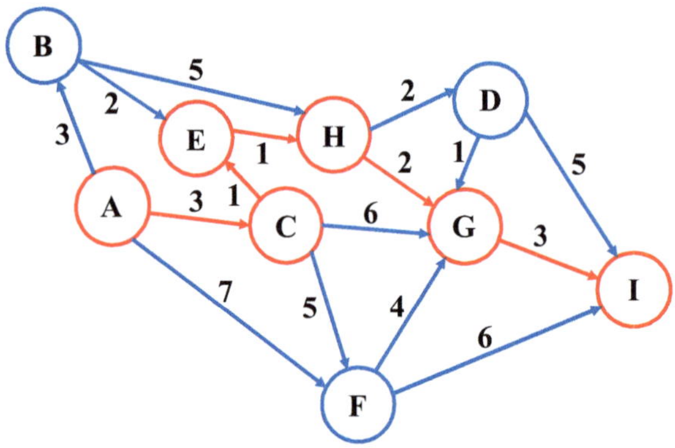

Fig. 5.2 The shortest pathExample 5.2

Only the final vertex remains in the set of undiscovered nodes. This vertex is then moved to the set of discovered nodes, and the forward pass is finished.

Now, let the backward pass is performed.

$$j_1 = p_9 = 7 \to j_2 = p_7 = 8 \to j_3 = p_8 = 5 \to j_4 = p_5 = 3 \to j_5 = p_3 = 1$$

That is, the shortest path from vertex 1 to 9 goes via nodes 3, 5, 8, and 7, and its length equals 10 (see Fig. 5.2).

It is obvious from Algorithm 5.2 that no vertex is passed more than once, while only its neighbors are searched only. The number of all possible paths is usually asymptotically much less than the number of all nodes.

References

1. Chong, E.K.P., Lu, W.S., Zak, S.H.: An Introduction to Optimization, 5th edn. Wiley, New York (2023)
2. Dijkstra, E.W.: A note on two problems in connexion with graphs. Numer. Math. **1**, 269–271 (1959). https://doi.org/10.1145/3544585.3544600
3. Fischetti, M.: Introduction to Mathematical Optimization. Independently Published (2019)
4. Fletcher, R.: Practical Methods of Optimization, 2nd edn. Wiley, New York (1987)
5. French, M.: Fundamentals of Optimization: Methods, Minimum Principles, and Applications for Making Things Better. Springer, Cham (2018)
6. Gill, P.E., Murray, W., Wright, M.H.: Practical Optimization. Academic Press, London (1981)
7. Guenin, B., Könemann, J., Tunçel, L.: A Gentle Introduction to Optimization. Cambridge University Press, Cambridge (2014)
8. Hillier, F.S., Lieberman, G.J.: Introduction to Operations Research, 7th edn. McGraw-Hill, New York (2002)
9. Lange, K.: Optimization, 2nd edn. Springer, New York (2013)
10. Luenberger, D.G., Ye, Y.: Linear and Nonlinear Programming, 3rd edn. Springer, New York (2008)
11. Rardin, R.L.: Optimization in Operations Research, 2nd edn. Pearson, London (2016)

12. Russell, S., Norvig, P.: Artificial Intelligence: A Modern Approach, 3rd edn. Prentice Hall, New York (2009)
13. Schrijver, A.: On the history of the shortest path problem. In: Optimization Stories (21st International Symposium on Mathematical Programming), Berlin, 2012. https://doi.org/10.4171/dms/6/19
14. Sniedovich, M.: Dynamic Programming: Foundations and Principles, 2nd edn. Francis & Taylor (CRC Press), Boca Raton (2010)
15. Tabak, D., Kuo, B.C.: Optimal Control by Mathematical Programming. Prentice Hall, New York (1971)
16. Walsh, G.R.: Methods of Optimization. Wiley, New York (1975)
17. Winston, W.L.: Operations Research: Applications and Algorithms. Brooks/Cole—Thomson Learning, Belmond (2004)

Chapter 6
Introduction to Decision and Game Theory

So far, this book has been devoted to formulating and solving problems in which a single decision-maker acted for himself/herself, and the values of the objective function have been clearly given. In everyday life, however, people often encounter tasks where they have to decide between several options, even though they do not know the exact consequences of their decision. Various random variables can affect this result (e.g., you are planning a birthday party and do not know how much food and drinks to buy because you cannot estimate which of the invited people will get sick and which will refuse to come for other reasons). In some cases, it is possible to estimate the approximate probability that a given result will occur (e.g., you know the weather forecast when planning a weekend trip with the family); however, the consequence of a given decision is completely random sometimes (e.g., which lottery ticket to buy from the offered pieces) [14]. A different type of problem occurs when another "intelligent" participant, typically a person or persons, enters the decision-making process. This type of *decision-making* process is called a *game* in this chapter [2, 3, 9, 11–13]. You can either compete with this other participant(s) or cooperate with him/her/them to achieve the greatest possible common benefit.

Basic decision-making situations and approaches to them, formal representations of the game, and the optimal solution to a conflict between two players who have a limited number of options for deciding (so-called strategies) are described in this chapter. The resolution of these conflicts is based on a mathematical model, with the help of which it is possible to obtain an optimal solution. The universality of these models is used in economics, logic, informatics, and other scientific branches. Specifically, matrix games are described here, representing the simplest form of a game between two *players*, where they compete with each other, and the consequences of *strategy* choices are presented in a matrix. Such games are also called constant-sum games.

In this chapter, the author of the book acknowledges his inspiration mainly from two books written in Czech [9, 11], in which references to many other sources in

L. Pekař, *Optimization: An Introduction*, Studies in Systems, Decision and Control 239,
https://doi.org/10.1007/978-3-031-86326-4_6

English can be found. Readers can also find the basics of decision and game theory in other publications that generally deal with parametric optimization [1, 4–8, 15, 16].

6.1 History and Preliminaries

The history of optimal decision-making and game theory as separate scientific disciplines is very young, but their basic ideas were formulated already in the distant past. A game, being a competition of two or more players, can date back to Antiques or even earlier. Chess, board games, card games, or cubes can be named as notorious examples. A human being was faced with decisions and behavior choices affected not only by other humans but also by the unpredictable effects of nature. He or she had to analyze the current situation and estimate the impacts of their decisions.

The decision and game theory—as a separate mathematically logical field of science—appeared in the eighteenth and nineteenth centuries. To name just a few scientists, an expected *utility* was formulated by Daniel Bernoulli (1700–1782) when solving the so-called St. Petersburg's paradox[1] in 1738. A general game theory (without in-depth mathematics) was introduced by Émile Borel (1871–1956), who studied a two-player game in which the players do not cooperate and have a finite number of possible strategies. The founder of mathematical game theory was John von Neumann (1903–1957), who proved the *fundamental theorem of game theory* in 1928 and published in 1944 together with Oskar Morgenstern (1902–1977) under the title "Game Theory and Economic Behavior". Another significant contribution to game theory was made by John Forbes Nash (1928–2015), who defined, e.g., the notion of the *equilibrium point* in the theory of non-cooperative games and won the Nobel Prize together with John Charles Harsanyi (1920–2000) and Reinhard Justus Reginald Selten (1930–2016) for economics. Even in 2005, the Nobel Prize for Economics was awarded to Thomas Crombie Schelling (1921–2016) and Robert John Aumann (1930) for the use of game theory.

Decision and game theories deal with *conflict decision situations*. However, problems for both theories should be distinguished and classified for the sake of this chapter.

For the decision theory, an intelligent decision maker is considered who is faced with a "non-intelligent" *random* decision maker. Three different cases can occur:

1. Decision under **certainty**
2. Decision under **risk**
3. Decision under **uncertainty**.

The first case represents a task when the decision maker deterministically knows the impact of their actions. In fact, this scenario coincides with the optimization

---

[1] A game where a coin is tossed until a "head" falls, and the reward is doubled with each roll. The mean value of the win is theoretically infinite but with an infinitesimally small but non-zero probability.

problems presented in the preceding chapters. Hence, this option is omitted in this chapter.

The decision under risk means that the intelligent decision maker knows the *probabilities* of the possible impacts of their particular decisions. Contrariwise, these probabilities are unknown when deciding under uncertainty.

Two or more *intelligent players* are assumed for game theory. The notion of "intelligent" has an alternative meaning of a "rational". It means that all decision-making options (the so-called *strategies*) and their outcomes are justified solely in rational (deductive and/or computational) operations. A rational decision is considered to be that for which all the reasons and consequences can be logically and consistently explained to other reasonable human beings. Nevertheless, a partially intelligent player can appear in a game. Such a player can do a rational decision; however, this decision is mostly non-optimal.

The basic classification of game types can be, for example, the following:

1. **Two-player** game
   (a) Adversarial (zero-sum) game
   (b) Non-adversarial game
      (i) Cooperative game
      (ii) Non-cooperative game
         (1) Shared payoff
         (2) Non-share payoff
2. **Multiple-player** game
   (a) Cooperative game
   (b) Non-cooperative game
      (1) Shared payoff
      (2) Non-share payoff.

The **adversarial** or **zero-sum** games are characterized by achieving the goal by one of the players, preventing the positive results of the others. The success of each player is possible only at the expense of the success of the other players. Contrariwise, all players have the opportunity to realize their goals in non-adversarial games. The *payoffs* of the players are not contradictory, so the players can also cooperate. When cooperating, the players can form *alliances* (if it is advantageous for them), in which they act not only according to their own goals but by the interests of the alliance. On the contrary, each player follows only their own goal in a non-cooperative game; however, it might be beneficial to share a payoff among the players if one has a significantly better position in a game.

In this introductory book, a specific type of the two-player zero-sum game is presented. This task is enjoyable in practice and represents a good starting point for studying other game types.

6.2 Decision Making

Basic notions and their mathematical representation from decision theory are concisely introduced. Then, two type of **decision-making situations** (or **processes**) and their potential solutions are presented.

A decision-making situation naturally includes its **participants**, the set of which is $Q = \{1, 2, \ldots, n\}$. Each participant has a finite or infinite **set of possible decisions**, i.e., $D = D_1 \times D_2 \times \cdots \times D_n$ are all these non-empty sets. And each participant ($i \in Q$) *selects* a **decision** (or **action**) from their set of possible decisions, i.e., $d_i \in D_i$, $i = 1, 2, \ldots, n$. There also exist *rules* and possibilities for how the participants can select their decisions and what are the corresponding possible **states** governed by the sets $X = \{x_1, x_2, \ldots, x_m\}$. The decision affects not only the participant who made the selection but also other participants. A particular **state** $x \in X$ might not be unambiguous; hence, the **outcome function** $f : D \to \phi(X)$ is related to each decision made, i.e., $f(d_1, d_2, \ldots, d_n) \in \phi(X)$, where $\phi(X)$ represents an operation with set X. Usually, each state $x_j \in X$ has a known or unknown probability of its occurrence.

However, the outcomes are perceived subjectively by individual participants. An acceptable or even optimal (i.e., the best possible) solution for one participant might not be acceptable for another one. Moreover, the outcomes have to be evaluated as related to other possible outcomes within the current state. Therefore, the expected utility function $u : D \to \mathbb{R}$ must be defined. The objective of each participant is to maximize the expected utility.

In the virtue of the above-introduced classification of the decision situations with *one intelligent participant* against random effects, the goal of the participant is to solve the problem

$$d^* = \arg \max_{d_i \in D} u(d_i) \tag{6.1}$$

where $D = D_1$. The following outcome functions can generally be defined (for each $d \in D$) for the decisions under certainty, risk, and uncertainty:

$$\begin{aligned} f_c(d) &= x \in X \\ f_r(d) &= \{[x, p_d(x)]\} \\ f_u(d) &= X_d \subseteq X \end{aligned} \tag{6.2}$$

respectively, where x means the corresponding unambiguous state, p_d expresses a probability distribution on X related to the particular decision, and X_d is a specific subset of possible states. That is, $\{[x, p_d(x)]\}$ in (6.2) expresses all the pairs: a state versus its probability, with $\sum_{x \in X} p_d(x) = 1$.

Decision situations under certainty and risk and several approaches for constructing related (expected) utility functions are introduced in the following two sections.

6.2.1 Decision Under Risk

This type of the decision process can be characterized by a *table* (or a matrix), the rows of which expresses the decisions $d_i \in D$ and its columns are possible states $x_j \in X$. Each state x_j can occur with probability $p_i(x_j)$, and $\sum_{x_j \in X} p_i(x_j) = 1$. The intersection of the selected d_i and the particular x_j (i.e., the table entry) gives rise to the outcome value $F_{i,j} = F(d_i, x_j) \in \mathbb{R}$ that expresses the state numerically. The situation is displayed in Table 6.1.

The question is how to construct the utility function $u(d)$ from $F_{\cdot,j}$ and $p_{\cdot}(x_j)$.

The **expected mean value** represents the simplest principle rule:

$$u(d_i) = E(F_{i,\cdot}) = \sum_{j=1}^{m} F_{i,j} p_i(x_j) \tag{6.3}$$

where $E(\cdot)$ means the mean value function.

Besides the mean value, a decision-situation participant can also be interested in the **variance**:

$$V(F_{i,\cdot}) = \sum_{j=1}^{m} (F_{i,j} - E(F_{i,j}))^2 p_i(x_j) \tag{6.4}$$

The goal is to maximize $E(F_{i,\cdot})$ and simultaneously minimize $V(F_{i,\cdot})$. Hence, if there exists a decision $d_i \in D$ so that

$$\begin{gathered} E(F_{i,\cdot}) > E(F_{k,\cdot}) \text{ and } V(F_{i,\cdot}) < V(F_{k,\cdot}) \\ \forall k = 1, 2, \ldots, n, \ k \neq i \end{gathered} \tag{6.5}$$

then d_i *dominates* the remaining possible decisions and is unambiguously preferred (i.e., the optimal) decision. If (6.5) does not hold, it is possible to alter expected utility function (6.3), e.g., as follows:

$$u(d_i) = E(F_{i,\cdot}) - \sqrt{V(F_{i,\cdot})} \tag{6.6}$$

Table 6.1 Decision under risk—the decision-situation table

	$[x_1, p(x_1)]$	$[x_2, p(x_2)]$	$\ldots$	$[x_{m-1}, p(x_{m-1})]$	$[x_m, p(x_m)]$
d_1	$F_{1,1}$	$F_{1,2}$	$\ldots$	$F_{1,m-1}$	$F_{1,m}$
d_2	$F_{2,1}$	$F_{2,2}$	$\ldots$	$F_{2,m-1}$	$F_{2,m}$
$\vdots$	$\vdots$	$\vdots$	$\ldots$	$\vdots$	$\vdots$
d_{n-1}	$F_{n-1,1}$	$F_{n-1,2}$	$\ldots$	$F_{n-1,m-1}$	$F_{n-1,m}$
d_n	$F_{n,1}$	$F_{n,2}$	$\ldots$	$F_{n,m-1}$	$F_{n,m}$

where $Std(F_{i,\cdot}) = \sqrt{V(F_{i,\cdot})}$ is the particular *standard deviation.*

However, many other possible definitions of $u(d_i)$ can be formulated, e.g., one can take normalized mean values and standard deviations in (6.6), etc.

Example 6.1 *Task*: Frankie wants to participate in the Monte Carlo Rally in a week.

Since he does not have a contract with a tire manufacturer, the choice of tires is up to him. He is considering the following brands: Mochalin, Purallo, and Bestyear. Each brand offers three types: slick, wet, and snow. The prestigious Pneu magazine has given the brands and types, on different surfaces, a rating of 0–100%, where 100% means perfect properties. This rating is shown in the following table:

	Dry surface	Wet surface	Snowy surface	Icy surface
Mochalin—slick	90	55	25	5
Mochalin—wet	75	85	45	15
Mochalin—snow	40	70	80	55
Purallo—slick	85	60	20	10
Purallo—wet	65	95	30	15
Purallo—snow	25	75	85	45
Bestyear—slick	80	60	25	15
Bestyear—wet	65	90	40	25
Bestyear—snow	10	50	95	65

The weather forecast is as follows: 20% dry, 40% wet, 30% snow, 10% ice.

According to principles (6.3) and (6.6), which racing car "shoes" should Frakie choose?

Solution: The participant (Frankie) has nine decision possibilities, which can be expressed by $D = \{1, 2, 3, 4, 5, 6, 7, 8, 9\}$ where the particular number agrees with the row number in the table. There are four states here: $X = \{\text{dry surface, wet surface, snowy surface, icy } \textit{surface}\} = \{1, 2, 3, 4\}$, which agree with the column numbers. Values $F_{i,j}$, $i = 1, 2, \ldots, 9$, $j = 1, 2, 3, 4$, are clear from the table. Probabilities of states corresponding to their number read $p(1) = 0.2, p(2) = 0.4, p(3) = 0.3, p(4) = 0.1$.

Follow rule (6.3) first. Calculate the mean value for each possible decision:

$$u(d_1) = E(F_{1,\cdot}) = 90 \cdot 0.2 + 55 \cdot 0.4 + 25 \cdot 0.3 + 5 \cdot 0.1 = 48$$
$$u(d_2) = E(F_{2,\cdot}) = 75 \cdot 0.2 + 85 \cdot 0.4 + 45 \cdot 0.3 + 15 \cdot 0.1 = 64$$
$$u(d_3) = E(F_{3,\cdot}) = 40 \cdot 0.2 + 70 \cdot 0.4 + 80 \cdot 0.3 + 55 \cdot 0.1 = 65.5$$
$$u(d_4) = E(F_{4,\cdot}) = 85 \cdot 0.2 + 60 \cdot 0.4 + 20 \cdot 0.3 + 10 \cdot 0.1 = 48$$
$$u(d_5) = E(F_{5,\cdot}) = 65 \cdot 0.2 + 95 \cdot 0.4 + 30 \cdot 0.3 + 15 \cdot 0.1 = 61.5$$
$$u(d_6) = E(F_{6,\cdot}) = 25 \cdot 0.2 + 75 \cdot 0.4 + 85 \cdot 0.3 + 45 \cdot 0.1 = 65$$
$$u(d_7) = E(F_{7,\cdot}) = 80 \cdot 0.2 + 60 \cdot 0.4 + 25 \cdot 0.3 + 15 \cdot 0.1 = 49$$
$$u(d_8) = E(F_{8,\cdot}) = 65 \cdot 0.2 + 90 \cdot 0.4 + 40 \cdot 0.3 + 25 \cdot 0.1 = 63.5$$
$$u(d_9) = E(F_{9,\cdot}) = 10 \cdot 0.2 + 50 \cdot 0.4 + 95 \cdot 0.3 + 65 \cdot 0.1 = 57$$

Hence, the "Mochalin—snow" type (i.e., $d^* = d_3$) gives the highest mean value and represents the best choice for Frankie. Notice that there is a higher chance of a wet surface, so tires made for a snowy surface give the best overall expected performance. Moreover, this type of tires does not have the best performance either on a wet surface or snow from all the possible types.

Now, try to compare this result with the optimal decision as per (6.6) where the expected variance is also considered:

$$u(d_1) = E(F_{1,\cdot}) - Std(F_{1,\cdot}) = 48 - 26.76 = 21.24$$
$$u(d_2) = E(F_{2,\cdot}) - Std(F_{2,\cdot}) = 64 - 28.37 = 35.63$$
$$u(d_3) = E(F_{3,\cdot}) - Std(F_{3,\cdot}) = 65.5 - 22.77 = 42.73$$
$$u(d_4) = E(F_{4,\cdot}) - Std(F_{4,\cdot}) = 48 - 26.66 = 21.34$$
$$u(d_5) = E(F_{5,\cdot}) - Std(F_{5,\cdot}) = 61.5 - 33.87 = 27.63$$
$$u(d_6) = E(F_{6,\cdot}) - Std(F_{6,\cdot}) = 65 - 28.44 = 36.56$$
$$u(d_7) = E(F_{7,\cdot}) - Std(F_{7,\cdot}) = 49 - 23.02 = 25.98$$
$$u(d_8) = E(F_{8,\cdot}) - Std(F_{8,\cdot}) = 63.5 - 28.91 = 34.59$$
$$u(d_9) = E(F_{9,\cdot}) - Std(F_{9,\cdot}) = 57 - 31.34 = 25.66$$

The optimal decision remains the same. What more! This choice dominates the remaining decisions, as it also has the minimum variance (and the standard deviation).

It is also worth noting that whereas d_6 ("Purallo—snow") has a utility function value very close to the optimal $d^* = d_3$ for (6.3), the distance between these two results becomes significantly higher for (6.6). Moreover, notice that $u(d_7) > u(d_6)$ holds for (6.3); however, it is $u(d_6) > u(d_7)$ for (6.6). Note that normalized functions in (6.6) would yield a slightly different order of best choices; however, the first four positions would remain the same.

6.2.2 Decision Under Uncertainties

This type of decision situation differs from that introduced in Sect. 6.2.1 in that the *probabilities* of particular states *are unknown*. Hence, it is possible to adopt the notation without considering $p_i(x_j)$; or equivalently, $p_i(x_j) = 1/m$ where $j = 1, 2, \ldots, m$. However, specific decision-making procedures (rules) and corresponding utility functions were proposed for the decision process under uncertainties. Let some of them be presented in this section.

The **minimax** (pessimistic) **principle** is searching for the worst states of each decision made first. Then, the maximum of these worst cases is taken as the optimal decision. Intuitively, one can consider the principle as the "lesser evil". It can mathematically be expressed as

$$\begin{aligned} u(d_i) &= \min F_{i,j} \\ j &= 1, 2, \ldots, m \end{aligned} \tag{6.7}$$

and

$$d^* = \arg\max_{d_i \in D} u(d_i) \tag{6.8}$$

On the contrary, the **maximax** (optimistic) **principle** takes the "best of the best", expecting the most positive state to occur. It can be expressed by

$$\begin{aligned} u(d_i) &= \max F_{i,j} \\ j &= 1, 2, \ldots, m \end{aligned} \tag{6.9}$$

followed by (6.8) again.

The **Hurwicz criterion** combines both the preceding principles and enables a weight on each of them. That is

$$\begin{aligned} u(d_i) &= \alpha \max F_{i,j} + (1 - \alpha) \min F_{i,j} \\ j &= 1, 2, \ldots, m \end{aligned} \tag{6.10}$$

where $\alpha \in [0, 1]$ represents the weight. Clearly, if $\alpha = 0$, the pessimistic principle is obtained, whereas the optimistic one takes $\alpha = 1$. Then, the optimal decision rule is given by (6.8).

The **minimax regret** rule is based on the observation that the selected decision is evaluated *ex-post*, i.e., after it is made. It minimizes the maximum differences in outcome values for particular decisions from the maximum outcome value and is helpful for risk-neutral participants. First, Table 6.1 is transformed into the *table of regrets* with entries

$$R_{i,j} = \max_{k=1,2,\ldots,n} F_{k,j} - F_{i,j}$$
$$i = 1, 2, \ldots, n \tag{6.11}$$
$$j = 1, 2, \ldots, m$$

Then, the maximum potential regrets are minimized

$$d^* = \arg\min_{d_i \in D}\left(\max_{j=1,2,\ldots,m} R_{i,j}\right) \tag{6.12}$$

Last but not least, the **Laplace principle** simply uses the consideration that all the states can occur with equal probability. Hence, the averaging rule (6.3) is applied with $p_i(x_j) = 1/m$.

Example 6.2 *Task*: Consider the problem formulated in Example 6.1; however, let Frankie make the tire selection a month earlier. Hence, he does not know the weather forecast. Compare the results obtained by the maximin and maximax principles, the Hurwicz criterion, the minimax regret rule, and the Laplace principle.

Solution: The maximin principle needs to calculate the worst cases for each possible state, see (6.7).

$$u(d_1) = \min\{90, 55, 25, 5\} = 5$$
$$u(d_2) = \min\{75, 85, 45, 15\} = 15$$
$$u(d_3) = \min\{40, 70, 80, 55\} = 40$$
$$u(d_4) = \min\{85, 60, 20, 10\} = 10$$
$$u(d_5) = \min\{65, 95, 30, 15\} = 15$$
$$u(d_6) = \min\{25, 75, 85, 45\} = 25$$
$$u(d_7) = \min\{80, 60, 25, 15\} = 15$$
$$u(d_8) = \min\{65, 90, 40, 25\} = 25$$
$$u(d_9) = \min\{10, 50, 95, 65\} = 10$$

Then, the maximum of these cases is found as per (6.8)

$$u^* = \max\{5, 15, 40, 10, 15, 25, 15, 25, 10\} = 40$$
$$\Rightarrow d^* = d_3$$

The optimistic (maximax) principle believes that the best state for any decision occurs, i.e.,

$$d^* = \arg\max\{90, 85, 80, 85, 95, 85, 80, 90, 95\} = \arg\{95\}$$
$$\Rightarrow d^* = d_5 \text{ or } d_9$$

Notice that d_3 (i.e., the best option so far) represents even the worst "optimistic" case, and the optimal solution is unambiguous.

Investigate the Hurwicz criterion for the trade-off setting $\alpha = 0.5$ in (6.10) now. One can observe that

$$u(d_1) = 0.5(90 + 5) = 47.5$$
$$u(d_2) = 0.5(85 + 15) = 50$$
$$u(d_3) = 0.5(80 + 40) = 60$$
$$u(d_4) = 0.5(85 + 10) = 47.5$$
$$u(d_5) = 0.5(95 + 15) = 55$$
$$u(d_6) = 0.5(85 + 25) = 55$$
$$u(d_7) = 0.5(80 + 15) = 47.5$$
$$u(d_8) = 0.5(90 + 25) = 57.5$$
$$u(d_9) = 0.5(95 + 10) = 52.5$$

Decision d_3 is optimal again since

$$d^* = \arg\max\{47.5, 50, 60, 47.5, 55, 55, 47.5, 57.5, 52.5\} = 60 = d_3$$

However, it can be calculated that for, e.g., $\alpha = 0.75$, the optimal decision reads $d^* = d_5$ with $u^* = u(d_5) = 75$.

The maximax regret rule represents the most complex decision strategy when making decision with uncertainties in this section. The table of regrets as per (6.11) has to be computed first:

	Dry surface	Wet surface	Snowy surface	Icy surface
Mochalin—slick	0	40	70	60
Mochalin—wet	15	10	50	50
Mochalin—snow	50	25	15	10
Purallo—slick	5	35	75	55
Purallo—wet	25	0	65	50
Purallo—snow	65	20	10	20
Bestyear—slick	10	35	70	50
Bestyear—wet	25	5	55	40
Bestyear—snow	80	45	0	0

Now, calculate a maximum regret for each possible decision, then select the minimum of these maxima. That is,

$$u^* = \min\{70, 50, 50, 75, 65, 65, 70, 55, 80\} = 50$$

which gives rise to

$$d^* = d_2 \text{ or } d_3$$

i.e., "Mochalin—wet" or "Mochalin—snow".

Finally, the Laplace principle simply computes all states with the equal probability of ¼, i.e.,

$$\begin{aligned}
u(d_1) &= 0.25(90 + 55 + 25 + 5) = 43.75\\
u(d_2) &= 0.25(75 + 85 + 45 + 15) = 55\\
u(d_3) &= 0.25(40 + 70 + 80 + 55) = 61.25\\
u(d_4) &= 0.25(85 + 60 + 20 + 10) = 43.75\\
u(d_5) &= 0.25(65 + 95 + 30 + 15) = 51.25\\
u(d_6) &= 0.25(25 + 75 + 85 + 45) = 57.5\\
u(d_7) &= 0.25(80 + 60 + 25 + 15) = 45\\
u(d_8) &= 0.25(65 + 90 + 40 + 25) = 55\\
u(d_9) &= 0.25(10 + 50 + 95 + 65) = 55
\end{aligned}$$

Hence, $d^* = d_3$ represents the best option again.

6.3 Games

In general, it is possible to adopt the concept of a decision process introduced at the beginning of Sect. 6.2. However, specific notions and notation for the **game** theory are used.

A game is played by **players** $Q = \{1, 2, \ldots, n\}$ (instead of participants). The players have their own **strategy spaces** $S_1, S_2, \ldots, S_n$ (instead of decisions). Each player $i \in Q$ selects a **strategy** $s_i \in S_i$, $i = 1, 2, \ldots, n$. These selections have an **outcome** (i.e., the impact on all the players). The outcomes are numerically represented by **payoff functions** $f_i : S_1 \times S_2 \times \cdots \times S_n \to \mathbb{R}$, $i = 1, 2, \ldots, n$ of each player (i.e., $f_i(s_1, s_2, \ldots, s_n) \in \mathbb{R}$). The goal of each player is to *maximize* their payoff.

Information is available to all players about all possible strategy spaces and payoff functions of both themselves and their opponents. Since players choose their strategies simultaneously, they have no way of knowing at the time of their decision which strategy their opponents will choose.

Mathematically, the game is characterized by the so-called *normal form game* model.

Definition 6.1 A **normal form game** is a set

$$\begin{gathered}
G := \{Q, S_i, f_i\}\\
i = 1, 2, \ldots, n
\end{gathered} \tag{6.13}$$

Symbols in Definition 6.1 are apparent from the text above.

Let the other two notions be defined to specify the game type that is further developed here.

Definition 6.2 A **finite game** is a game for which it holds that

$$
\begin{aligned}
&|S_i| < \infty \\
&\forall i = 1, 2, \ldots, n
\end{aligned}
\tag{6.14}
$$

Otherwise, the game is **infinite**.

Definition 6.3 A **constant-sum game** is a game, for which it holds that

$$
\sum_{i=1}^{n} f_i = c \tag{6.15}
$$

where c is a constant, regardless of the strategy selection. This kind of game is also called **adversarial**.

To rephrase Definition 6.2, all the strategy spaces have a finite number of strategies in a finite game. For example, a closed interval $S = [0, 1]$ represents an infinite strategy space, while a set $S = \{0, 1, 2\}$ is finite.

A constant-sum game can be imagined as a "fight" for one common source, characteristic for war conflicts, card or board games, etc. If one wins, another loses. A **zero-sum game** represents a special case of a constant-sum game which is characterized by $c = 0$.

6.3.1 2-Player Antagonistic Conflict—A Single-Matrix Game

Consider an adversarial conflict situation between two intelligent players with only a limited number of possible game strategies in this section, representing the simplest game type. In light of Definitions 6.2 and 6.3, it can be classified as a *finite constant-sum game*. The following statement holds for two-player constant-sum games.

Theorem 6.1 *Assume a two-player constant-sum game. This game is equivalent to a two-player zero-sum game, i.e., both have identical optimal strategies.*

Hence, $c = 0$ can be considered without loss of generality. The normal form of this game reads

$$
G = \left\{
\begin{array}{l}
Q = \{1, 2\}, S_1 = \{s_{11}, s_{12}, \ldots, s_{1N}\}, S_2 = \{s_{21}, s_{22}, \ldots, s_{2M}\}, \\
f_1(s_1, s_2) + f_2(s_1, s_2) = 0, s_1 \in S_1, s_2 \in S_2
\end{array}
\right\}
\tag{6.16}
$$

The crucial idea and guide on finding the **optimal strategies** $s_1^* \in S_1, s_2^* \in S_2$ is based on the following statement about the **equilibrium point** (also called the Nash equilibrium).

Theorem 6.2 *The optimal player's strategy is that changing the choice would not bring a better result, assuming that the opponent also selects their optimal strategy.*

In other words, if a player goes wrong and, simultaneously, their opponent chooses the optimal strategy, the player gets an equal or worse payoff. However, the same holds from the opponent's viewpoint. Theorem 6.2 can be mathematically formulated as follows:

$$\begin{aligned} f_1\left(s_1, s_2^*\right) &\le f_1\left(s_1^*, s_2^*\right) \\ f_2\left(s_1^*, s_2\right) &\le f_2\left(s_1^*, s_2^*\right) \end{aligned} \tag{6.17}$$

It can be shown that a common payoff for game (6.16) can be introduced:

$$f(s_1, s_2) = f_1(s_1, s_2) - f_2(s_1, s_2) \tag{6.18}$$

It means that player 1 attempts to maximize $f(\cdot)$, while player 2 needs to minimize the common payoff. If $f(\cdot) > 0$, player 1 wins; otherwise, whenever $f(\cdot) < 0$, player 2 wins. The equilibrium point (6.17) can be expressed using (6.18) as follows:

$$f\left(s_1, s_2^*\right) \le f\left(s_1^*, s_2^*\right) \le f\left(s_1^*, s_2\right) \tag{6.19}$$

Expression $f\left(s_1^*, s_2^*\right)$ in condition (6.19) geometrically means a saddle point of $f(s_1, s_2)$, and its value is called the **game cost**.

Model (6.16) by means of (6.19) can further be transformed into the so-called **game matrix**:

$$\mathbf{F} = \begin{bmatrix} f_{11} & f_{12} & \cdots & f_{1M} \\ f_{21} & f_{22} & \cdots & f_{2M} \\ \vdots & \vdots & \ddots & \vdots \\ f_{N1} & f_{N1} & \cdots & f_{NM} \end{bmatrix} \tag{6.20}$$

where $f_{ij} = f\left(s_{1i}, s_{2j}\right)$. That is, player 1 selects rows representing their possible strategies, while player 2 chooses columns that agree with their possible strategies. The former wants to find a maximum entry of **F**, whereas the latter searches for a minimum element. Hence, (6.16) becomes

$$G = \left\{ \begin{array}{l} Q = \{1, 2\}, S_1 = \{1, 2, \ldots, N\}, S_2 = \{1, 2, \ldots, M\}, \mathbf{F}(i, j) = f_{ij}, \\ i \in S_1, j \in S_2 \end{array} \right\} \tag{6.21}$$

Game (6.21) can either have a unique (unambiguous) optimal solution called the **pure strategy** or, otherwise, a **mixed strategy**.

6.3.1.1 Pure Strategy

A pure strategy means that whenever the game repeats, the player still plays the same optimal strategy since they could only make themselves worse by choosing a different strategy (provided that the opponent also chooses only the optimal strategy). In accordance with (6.19), the pure strategy exists if matrix **F** includes its entry that is a *maximum* in a *column*, and, simultaneously, a *minimum* in a particular *row*. This entry is called a **saddle element** of **F**, which mathematically be expressed as follows. The aim of player 1 is to find

$$ {}^{1}f_{ij}^{*} = \max_{i} \min_{j} f_{ij} \tag{6.22}$$

for $i = 1, 2, \ldots, N$, $j = 1, 2, \ldots, M$, while the goal of player 2 reads

$$ {}^{2}f_{ij}^{*} = \min_{j} \max_{i} f_{ij} \tag{6.23}$$

Value ${}^{1}f_{ij}^{*}$ represents the *lower* game cost (i.e., it maximizes the lowest possible wins of player 1), whereas ${}^{2}f_{ij}^{*}$ means the *upper* game cost (i.e., it minimizes the highest possible losses of player 2).

The saddle element of **F**, and hence, the (optimal) pure strategy $\{i^*, j^*\}$ exists if and only if

$$ {}^{1}f_{ij}^{*} = {}^{2}f_{ij}^{*} = f^{*} \tag{6.24}$$

where $\{i^*, j^*\} = \{i, j\}$.

Example 6.3 *Task*: Johnny and Frankie are high school classmates who are interested in two girls, Mary and Lilly. Mary is prettier and smarter, that is why both guys value her 90%. Lilly is also beautiful, but not up to Mary's level (as she cannot dance). Hence, both boys rate her at 60%. Johnny and Frankie do not know which one to court and how much energy to sacrifice. Simultaneously, both girls are thinking about these boys.

Each guy has the energy to either fully attract one girl or only partially attract each girl. If only one of the boys is interested in the girl, he will have her all to himself. If both boys are thoroughly interested in her, her favor is split between them equally. Finally, if one guy is fully interested in one girl and the other is partially interested, the former will get twice the feelings of that girl than the one who is not 100% interested.

Find the optimal strategy of both boys following principle (6.19).

Solution: Each boy has three possible strategies: $S_1 = S_2 = \{1, 2, 3\}$. Let strategy 1 be "courting Mary fully", strategy 2 "courting Lilly fully", and strategy 3 "courting both girls equally". Analyze the payoffs of each player (boy).

For instance, if $s_1 = 1$ and also $s_2 = 1$, both guys want to have Mary, and they get 50% of her interest, i.e., a rating of 45% on the overall beauty scale. Simultaneously, Lilly does not know about the endeavors of both boys; therefore, the guys also get 50% of her interest, i.e., a rating of 30% on the overall beauty scale. That is, $f_1(1,1) = f_2(1,1) = 45 + 30 = 75$ and $f_{11} = f_1(1,1) - f_2(1,1) = 0$ from (6.18).

If $s_1 = 1$ and $s_2 = 2$, each boy attracts his girl; therefore, $f_1(1,2) = 90$ and $f_2(1,2) = 60$. Hence, $f_{12} = f_1(1,2) - f_2(1,2) = 30$. As a final detailed example, considering $s_1 = 1$ and $s_2 = 3$, Mary prefers Johnny 2:1 compared to Frankie, i.e., a rating of 60% on the overall beauty scale goes to Johnny and 30% goes to Frankie. Simultaneously, only Frankie is courting Lilly; therefore, he will fully attract her (as she does not know about his feelings for Mary) and earn 60% of the overall beauty scale in addition. To sum up, $f_1(1,3) = 60$ and $f_2(1,3) = 30 + 60 = 90$, which gives rise to $f_{13} = f_1(1,3) - f_2(1,3) = -30$.

Further analyses yield:

$$\begin{aligned}
&f_1(2,1) = 60, f_2(2,1) = 90 \Rightarrow f_{21} = 60 - 90 = -30\\
&f_1(2,2) = f_2(2,2) = 45 + 30 = 75 \Rightarrow f_{22} = 75 - 75 = 0\\
&f_1(2,3) = 40, f_2(2,3) = 90 + 20 = 110 \Rightarrow f_{23} = 40 - 110 = -70\\
&f_1(3,1) = 30 + 60 = 90, f_2(3,1) = 60 \Rightarrow f_{31} = 90 - 60 = 30\\
&f_1(3,2) = 90 + 20 = 110, f_2(3,2) = 40 \Rightarrow f_{32} = 110 - 40 = 70\\
&f_1(3,3) = f_2(3,3) = 45 + 30 = 75 \Rightarrow f_{33} = 75 - 75 = 0
\end{aligned}$$

Then, the game matrix reads

$$\mathbf{F} = \begin{bmatrix} 0 & 30 & -30 \\ -30 & 0 & -70 \\ 30 & 70 & 0 \end{bmatrix}$$

Following (6.22) and (6.23), one gets

$$\begin{aligned}
{}^1f_{ij}^* &= \max_i\{-30, -70, 0\} = 0 \Rightarrow \{i,j\} = \{3,3\}\\
{}^2f_{ij}^* &= \min_j\{30, 70, 0\} = 0 \Rightarrow \{i,j\} = \{3,3\}
\end{aligned}$$

i.e., ${}^1f_{ij}^* = {}^2f_{ij}^* = f^* = 0$ with $\{i^*, j^*\} = \{3,3\}$, which represents the saddle element of $\mathbf{F}$.

Hence, the optimal strategy for both boys is to court both girls equally, and each gets half of their beauty.

Two interesting and important notions relate to pure strategies and Example 6.3.

Definition 6.4 A strategy $i \in S_1$ is **dominating** if

$$\forall k = 1, 2, \ldots, N \neq i : f_{ij} \geq f_{kj}, \forall j = 1, 2, \ldots, M \tag{6.25}$$

A strategy $i \in S_1$ is **dominated** if

$$\exists k = 1, 2, \ldots, N \neq i : f_{ij} \leq f_{kj}, \forall j = 1, 2, \ldots, M \tag{6.26}$$

Note that analogous definitions can be made for player 2.

If a strategy is dominating, the particular row or column of **F** represents a *pure strategy* of a particular player. Then, the opponent takes the minimum element in the ith row (for player 2) or the maximum element in the jth column (for player 1).

If a strategy is dominated, the particular row or column can be *deleted* from the game matrix.

Definition 6.5 Game (6.21) is **symmetric** if

$$\mathbf{F} = -\mathbf{F}^T \tag{6.27}$$

where superscript T means the matrix transpose. That is, **F** is *antisymmetric*.

6.3.1.2 Mixed Strategy

Assume now that (6.24) does not hold. In this case, (optimal) *mixed* strategies of both players have to be found. Each strategy is given a probability. The sum of probabilities must equal one.

Hence, game normal form (6.21) is adjusted to

$$G = \left\{ \begin{array}{l} Q = \{1, 2\}, S_1 = \left\{ \mathbf{s}_1 = [s_{11}, s_{12}, \ldots, s_{1N}]^T : \sum_{i=1}^{N} s_{1i} = 1, s_{1i} \geq 0 \right\}, \\ S_2 = \left\{ \mathbf{s}_2 = [s_{21}, s_{22}, \ldots, s_{2M}]^T : \sum_{j=1}^{M} s_{2j} = 1, s_{2j} \geq 0 \right\}, \\ f(\mathbf{s}_1, \mathbf{s}_2) = \sum_{i=1}^{N} \sum_{j=1}^{M} \mathbf{s}_1^T f_{ij} \mathbf{s}_2, f_{ij} \in \mathbf{F} \end{array} \right\} \tag{6.28}$$

where $f(\mathbf{s}_1, \mathbf{s}_2)$ is the payoff function to be optimized in the sense of (6.19).

The following crucial theorem applies here.

Theorem 6.3 (The fundamental theorem of game theory [13]) *Let* $\mathbf{F} = [f_{ij}]$ *be the game (payoff) matrix of a two-player zero-sum game and* $\mathbf{s}_1 \in S_1, \mathbf{s}_2 \in S_2$*, strategies representing probabilities of pure strategies as in (6.28). The payoff function (i.e., expected winning) is given by* $f(\mathbf{s}_1, \mathbf{s}_2) = \sum_{i=1}^{N} \sum_{j=1}^{M} \mathbf{s}_1^T f_{ij} \mathbf{s}_2$*. Then, every matrix game has an optimal solution in mixed strategies, i.e., there always exist* $\mathbf{s}_1^*, \mathbf{s}_2^*$ *for which*

$$^1f_{ij}^* = \max_i \min_j f_{ij} = f^* = f\left(\mathbf{s}_1^*, \mathbf{s}_2^*\right) = \min_j \max_i f_{ij} = {}^2f_{ij}^* \tag{6.29}$$

where f^* is the optimal game cost.

It is worth noting that the correct interpretation of the optimal probability s_{ij}^* is as follows: If the game is played n-times where $n \to \infty$, the i th player selects j th strategy $s_{ij}^* n$-times.

The remaining question is how the optimal pair $\mathbf{s}_1^*, \mathbf{s}_2^*$ is calculated. Assume the following proposition first.

Proposition 6.1 *Consider game matrix $\mathbf{F} = \left[f_{ij}\right]$ with a payoff function value $f(\mathbf{s}_1, \mathbf{s}_2)$ for some fixed $\mathbf{s}_1 \in S_1, \mathbf{s}_2 \in S_2$. Then, game matrix $\mathbf{F}_\alpha = \left[f_{ij} + \alpha\right]$ (i.e., a number α is added to every element of $\mathbf{F}$) has a payoff function value of $f(\mathbf{s}_1, \mathbf{s}_2) + \alpha$ for the same strategies $\mathbf{s}_1 \in S_1, \mathbf{s}_2 \in S_2$.*

A guide for searching $\mathbf{s}_1^*, \mathbf{s}_2^*$ can be derived from the following observations. Assume that $\mathbf{F}$ includes only non-negative entries. It is known from (6.19) that $f\left(\mathbf{s}_1^*, \mathbf{s}_2^*\right) = f^* \leq f\left(\mathbf{s}_1^*, \mathbf{s}_2\right)$. It means that if player 1 keeps their optimal strategy $\mathbf{s}_1^* = \left[s_{11}^*, s_{12}^*, \ldots, s_{1N}^*\right]^T$ with $\sum_{i=1}^N s_{1i}^* = 1, s_{1i}^* \geq 0$, but player 2 does not, the game cost increases or remains the same. It also must hold for any pure strategy of player 2. Hence, if player 2 chooses the pure strategy $\mathbf{s}_2 = [1, 0, \ldots, 0]^T$ (i.e., "the first column of $\mathbf{F}$"), it holds

$$\sum_{i=1}^{N} f_{i1} s_{1i}^* = f_{11} s_{11}^* + f_{21} s_{12}^* + \cdots + f_{N1} s_{1N}^* \geq f_{ij}^* \tag{6.30}$$

Analogous conditions can be obtained for $\mathbf{s}_2 = [0, 1, 0, \ldots, 0]^T$, $\mathbf{s}_2 = [0, 0, 1, 0, \ldots, 0]^T$, etc. Then, the following set of inequalities is obtained

$$\begin{aligned}
&\sum_{i=1}^{N} f_{i1} s_{1i}^* = f_{11} s_{11}^* + f_{21} s_{12}^* + \cdots + f_{N1} s_{1N}^* \geq f^* \\
&\sum_{i=1}^{N} f_{i2} s_{1i}^* = f_{12} s_{11}^* + f_{22} s_{12}^* + \cdots + f_{N2} s_{1N}^* \geq f^* \\
&\vdots \\
&\sum_{i=1}^{N} f_{iM} s_{1i}^* = f_{1M} s_{11}^* + f_{2M} s_{12}^* + \cdots + f_{NM} s_{1N}^* \geq f^*
\end{aligned} \tag{6.31}$$

Introduce relative probabilities

$$\bar{\mathbf{s}}_1 = \mathbf{s}_1 / f^* \tag{6.32}$$

Then, (6.31) becomes

$$
\begin{aligned}
&f_{11}\bar{s}_{11}^{*} + f_{21}\bar{s}_{12}^{*} + \cdots + f_{N1}\bar{s}_{1N}^{*} \geq 1 \\
&f_{12}\bar{s}_{11}^{*} + f_{22}\bar{s}_{12}^{*} + \cdots + f_{N2}\bar{s}_{1N}^{*} \geq 1 \\
&\vdots \\
&f_{1M}\bar{s}_{11}^{*} + f_{2M}\bar{s}_{12}^{*} + \cdots + f_{NM}\bar{s}_{1N}^{*} \geq 1
\end{aligned}
\tag{6.33}
$$

with

$$\sum_{i=1}^{N} \bar{s}_{1i}^{*} = 1/f^{*} \tag{6.34}$$

Hence, the problem of searching for the optimal strategy can be transformed into an LP problem that minimizes the linear cost function

$$\varphi_{\min}(\bar{\mathbf{s}}_1) = \bar{s}_{11}` + \bar{s}_{12} + \cdots + \bar{s}_{1N} = \mathbf{1}^T \bar{\mathbf{s}}_1 \tag{6.35}$$

where $\mathbf{1} = [1, 1, \ldots, 1]^T$ (with a known relation between the cost function $\varphi_{\min}\left(\bar{\mathbf{s}}_1^*\right) = \varphi_{\min}^*$ and f^* given by (6.34)), subject to constraint set (6.33) in $\bar{\mathbf{s}}_1$ that can be expressed in a condensed form

$$\mathbf{F}^T \bar{\mathbf{s}}_1 \geq \mathbf{1} \tag{6.36}$$

Analogous derivations can be done from the viewpoint of player 2. Here, condition $f\left(\mathbf{s}_1^*, \mathbf{s}_2^*\right) = f^* \geq f\left(\mathbf{s}_1, \mathbf{s}_2^*\right)$ (when assuming pure strategies $\mathbf{s}_1 = [1, 0, \ldots, 0]^T$, $\mathbf{s}_1 = [0, 1, 0, \ldots, 0]^T$) yields

$$
\begin{aligned}
&\sum_{j=1}^{M} f_{1j}s_{2j}^{*} = f_{11}s_{21}^{*} + f_{12}s_{22}^{*} + \cdots + f_{1M}s_{2M}^{*} \leq f^{*} \\
&\sum_{j=1}^{M} f_{2j}s_{2j}^{*} = f_{12}s_{21}^{*} + f_{22}s_{22}^{*} + \cdots + f_{2M}s_{2M}^{*} \leq f^{*} \\
&\vdots \\
&\sum_{j=1}^{M} f_{Nj}s_{2j}^{*} = f_{N1}s_{21}^{*} + f_{N2}s_{22}^{*} + \cdots + f_{NM}s_{2M}^{*} \leq f^{*}
\end{aligned}
\tag{6.37}
$$

When defining relative probabilities

$$\bar{\mathbf{s}}_2 = \mathbf{s}_2 / f^* \tag{6.38}$$

set (6.37) is transformed into

$$
\begin{aligned}
f_{11}\bar{s}_{21}^{*} + f_{12}\bar{s}_{22}^{*} + \cdots + f_{1M}\bar{s}_{2M}^{*} &\leq 1 \\
f_{21}\bar{s}_{21}^{*} + f_{22}\bar{s}_{22}^{*} + \cdots + f_{2M}\bar{s}_{2M}^{*} &\leq 1 \\
&\vdots \\
f_{N1}\bar{s}_{21}^{*} + f_{N2}\bar{s}_{22}^{*} + \cdots + f_{NM}\bar{s}_{2M}^{*} &\leq 1
\end{aligned}
\tag{6.39}
$$

with

$$
\sum_{j=1}^{M} \bar{s}_{2j}^{*} = 1/f^{*} \tag{6.40}
$$

Hence, a combination of (6.38) and (6.39) gives rise to the following maximization LP problem

$$
\varphi_{\max}(\bar{\mathbf{s}}_2) = \bar{s}_{21} + \bar{s}_{22} + \cdots + \bar{s}_{2M} = \mathbf{1}\bar{\mathbf{s}}_2 \tag{6.41}
$$

subject to

$$
\mathbf{F}\bar{\mathbf{s}}_2 \leq \mathbf{1} \tag{6.42}
$$

with the knowledge of $\varphi_{\max}\left(\bar{\mathbf{s}}_2^{*}\right) = \varphi_{\max}^{*} = 1/f^{*}$.

Notice that $\varphi_{\min}\left(\bar{\mathbf{s}}_1^{*}\right) = \varphi_{\max}\left(\bar{\mathbf{s}}_2^{*}\right) = 1/f^{*}$, and LP problems (6.35)–(6.36) and (6.41)–(6.42) are mutually dual (see Sect. 4.2.4). Therefore, searching for the optimal mixed strategies for game (6.28) can be transformed into solving the primary LP problem (6.41)–(6.42) (e.g., using the simplex table— see Algorithm4.1) when considering the dual problem (6.35)–(6.36).

The following algorithm summarizes the particular solution steps.

Algorithm 6.1 (Optimal mixed strategies in a finite two-player zero-sum game)

1. Let the game matrix $\mathbf{F}$ be given. If $\exists f_{ij} < 0 \in \mathbf{F}$ for some $i = 1, 2, \ldots, N$ and $j = 1, 2, \ldots, M$, select $\alpha \geq -\min f_{ij}$ for all $f_{ij} < 0$; otherwise, set $\alpha = 0$.
2. Set $f_{\alpha,ij} = f_{ij} + \alpha$ for all $i = 1, 2, \ldots, N$ and $j = 1, 2, \ldots, M$, and assemble matrix $\mathbf{F}_{\alpha} = \left[f_{\alpha,ij}\right]$.
3. Solve the primary LP problem (6.41)–(6.42) for $\mathbf{F}_{\alpha}$ using Algorithm 4.1, which gives rise to optimal structural variables $\bar{\mathbf{s}}_2^{*}$ of the primal problem. Detect optimal values $\bar{\mathbf{s}}_1^{*}$ of the corresponding dual problem (6.35)–(6.36) as per Sect. 4.2.4. Determine also the optimal objective function value $\varphi_{\max}\left(\bar{\mathbf{s}}_2^{*}\right) = \varphi_{\max}^{*}$.
4. Calculate the actual optimal probabilities of particular strategies as $\mathbf{s}_1^{*} = \bar{\mathbf{s}}_1^{*}/\varphi_{\max}^{*}$ and $\mathbf{s}_2^{*} = \bar{\mathbf{s}}_2^{*}/\varphi_{\max}^{*}$.
5. The actual optimal game cost reads $f^{*} = 1/\varphi_{\max}^{*} - \alpha$.

Example 6.4 *Task*: Find optimal strategies for the "rock-paper-scissors" game of two players. That is, "paper" beats "rock", "rock" beats "scissors", and "scissors" beats "paper".

Solution: Let, for instance, a win be awarded with a gain of 1 point, a loss with a loss of 1 points, and a draw of 0 points. Assign the first strategy to "rock", the second to "paper" and the third to "scissors". Hence, one has

$$f_{11} = 0 - 0 = 0,\ f_{12} = 1 - (-1) = 2,\ f_{13} = -1 - 1 = -2$$
$$f_{21} = -1 - 1 = -2,\ f_{22} = 0 - 0 = 0,\ f_{23} = 1 - (-1) = 2$$
$$f_{31} = 1 - (-1) = 2,\ f_{32} = -1 - 1 = -2,\ f_{33} = 0 - 0 = 0$$

i.e.,

$$\mathbf{F} = \begin{bmatrix} 0 & 2 & -2 \\ -2 & 0 & 2 \\ 2 & -2 & 0 \end{bmatrix}$$

Apparently, the saddle element of **F** as per (6.22)–(6.24) does not exist. Note that the game is symmetric as **F** is antisymmetric (see Definition 6.5), which is natural.

Because $\min f_{ij} = -2$, select $\alpha = 2$, giving rise to

$$\mathbf{F}_\alpha = \begin{bmatrix} 2 & 4 & 0 \\ 0 & 2 & 4 \\ 4 & 0 & 2 \end{bmatrix}$$

The LP problem as per (6.41) and (6.42) reads

$$\max \varphi_{\max}(\overline{\mathbf{s}}_2) = \overline{s}_{21} + \overline{s}_{22} + \overline{s}_{23}$$
$$2\overline{s}_{21} + 4\overline{s}_{22} \leq 1$$
$$2\overline{s}_{22} + 4\overline{s}_{23} \leq 1$$
$$4\overline{s}_{21} + 2\overline{s}_{23} \leq 1$$

The corresponding simplex table has form

$\overline{s}_{21}$	$\overline{s}_{22}$	$\overline{s}_{23}$	$\overline{s}_{24}$	$\overline{s}_{25}$	$\overline{s}_{26}$	b_i
2	4	0	1	0	0	1
0	2	4	0	1	0	1
4	0	2	0	0	1	1
− 1	− 1	− 1	0	0	0	0

where $\overline{s}_{24}$, $\overline{s}_{25}$, and $\overline{s}_{26}$ are slack variables. Apparently, there are always as many possibilities to select the pivot columns as the possible strategies of player 2 are.

Without loss of generality, select the left-most column, determine the pivot element (highlighted in italics), and eliminate it:

$\bar{s}_{21}$	$\bar{s}_{22}$	$\bar{s}_{23}$	$\bar{s}_{24}$	$\bar{s}_{25}$	$\bar{s}_{26}$	b_i
0	*4*	− 1	1	0	− 0.5	0.5
0	2	4	0	1	0	1
1	0	1/2	0	0	0.25	0.25
0	− 1	− 0.5	0	0	0.25	0.25

The remaining steps of Algorithm 4.1 are:

$\bar{s}_{21}$	$\bar{s}_{22}$	$\bar{s}_{23}$	$\bar{s}_{24}$	$\bar{s}_{25}$	$\bar{s}_{26}$	b_i
0	1	− 0.25	0.25	0	− 0.125	0.125
0	0	*4.5*	− 0.5	1	0.25	0.75
1	0	0.5	0	0	0.25	0.25
0	0	− 0.75	0.25	0	0.125	0.375

$\bar{s}_{21}$	$\bar{s}_{22}$	$\bar{s}_{23}$	$\bar{s}_{24}$	$\bar{s}_{25}$	$\bar{s}_{26}$	b_i
0	1	0	2/9	1/18	− 1/9	1/6
0	0	1	− 1/9	2/9	1/18	1/6
1	0	0	1/18	− 1/9	2/9	1/6
0	0	0	1/6	1/6	1/6	0.5

Following step 3 of Algorithm 6.1 and considering the dual model, one gets

$$\bar{\mathbf{s}}_1^* = \left[\bar{s}_{11}^*, \bar{s}_{12}^*, \bar{s}_{13}^*\right]^T = [1/6, 1/6, 1/6]^T$$
$$\bar{\mathbf{s}}_2^* = \left[\bar{s}_{21}^*, \bar{s}_{22}^*, \bar{s}_{23}^*\right]^T = [1/6, 1/6, 1/6]^T$$
$$\varphi_{\max}^* = 0.5$$

Step 4 and 5 of Algorithm 6.1 dictate

$$\mathbf{s}_1^* = \frac{\bar{\mathbf{s}}_1^*}{\varphi_{\max}^*} = 2\left[\frac{1}{6}, \frac{1}{6}, \frac{1}{6}\right]^T = \left[\frac{1}{3}, \frac{1}{3}, \frac{1}{3}\right]^T$$
$$\mathbf{s}_2^* = \frac{\bar{\mathbf{s}}_2^*}{\varphi_{\max}^*} = 2\left[\frac{1}{6}, \frac{1}{6}, \frac{1}{6}\right]^T = \left[\frac{1}{3}, \frac{1}{3}, \frac{1}{3}\right]^T$$
$$f^* = \frac{1}{\varphi_{\max}^*} - \alpha = 2 - 2 = 0$$

Hence, the optimal strategy of both players of the "rock-paper-scissors" game is to choose each of the three possibilities equally. For instance, if there are 99 game rounds, each player should take "rock" 33 times, "paper" 33 times, and "scissors" 33 times (randomly).

The game cost is zero, which means that if the players behave optimally, the game is equally fair for both.

References

1. Chong, E.K.P., Lu, W.S., Zak, S.H.: An Introduction to Optimization, 5th edn. Wiley, New York (2023)
2. Clegg, B.: Game Theory: Understanding the Mathematics of Life. Icon Books, London (2020)
3. Ferguson, T.S.: A Course in Game Theory. World Industries Scientific Publishing, Singapore (2020)
4. Fletcher, R.: Practical Methods of Optimization, 2nd edn. Wiley, New York (1987)
5. French, M.: Fundamentals of Optimization: Methods, Minimum Principles, and Applications for Making Things Better. Springer, Cham (2018)
6. Gill, P.E., Murray, W., Wright, M.H.: Practical Optimization. Academic Press, London (1981)
7. Guenin, B., Könemann, J., Tunçel, L.: A Gentle Introduction to Optimization. Cambridge University Press, Cambridge (2014)
8. Hillier, F.S., Lieberman, G.J.: Introduction to Operations Research, 7th edn. McGraw-Hill, New York (2002)
9. Hykšová, M.: Teorie her a optimální rozhodování (Game Theory and Optimal Decision). ČVUT, Prague (2009)
10. Lange, K.: Optimization, 2nd edn. Springer, New York (2013)
11. Maňas, M.: Teorie her a optimální rozhodování (Game Theory and Optimal Decision). SNTL, Prague (1974)
12. Osborne, M.J., Rubinstein, A.: A Course in Game Theory. MIT Press, Cambridge (1994)
13. Panik, M.J.: The fundamental theorem of game theory revisited. Appl. Math. Lett. 7(2), 77–79 (1994)
14. Pekař, L., Matušů, R., Andrla, J., Litschmannová, M.: Review of Kalah game research and the proposition of a novel heuristic–deterministic algorithm compared to tree-search solutions and human decision-making. Informatics **7**, 34 (2020). https://doi.org/10.3390/informatics7030034
15. Peterson, M.: An Introduction to Decision Theory. Cambridge University Press, Cambridge (2009)
16. Rardin, R.L.: Optimization in Operations Research, 2nd edn. Pearson, London (2016)
17. Winston, W.L.: Operations Research: Applications and Algorithms. Brooks/Cole—Thomson Learning, Belmond (2004)

Index

L. Pekař, *Optimization: An Introduction*, Studies in Systems, Decision and Control 239,
https://doi.org/10.1007/978-3-031-86326-4

N

O

P

MIX
Papier aus verantwortungsvollen Quellen
Paper from responsible sources
FSC® C105338

If you have any concerns about our products,
you can contact us on
ProductSafety@springernature.com

In case Publisher is established outside the EU,
the EU authorized representative is:
Springer Nature Customer Service Center GmbH
Europaplatz 3, 69115 Heidelberg, Germany

Printed by Libri Plureos GmbH
in Hamburg, Germany